The Home Radio: How To Make and Use It

by A. Hyatt Verrill

with an introduction by Roger Chambers

This work contains material that was originally published in 1922.

This publication was created and published for the public benefit, utilizing public funding and is within the Public Domain.

This edition is reprinted for educational purposes and in accordance with all applicable Federal Laws.

Introduction Copyright 2018 by Roger Chambers

Self Reliance Books

Get more historic titles on animal and stock breeding, gardening and old fashioned skills by visiting us at:

http://selfreliancebooks.blogspot.com/

Introduction

I am pleased to present yet another title in our "How To ..." series.

The work is in the Public Domain and is re-printed here in accordance with Federal Laws.

As with all reprinted books of this age that are intended to perfectly reproduce the original edition, considerable pains and effort had to be undertaken to correct fading and sometimes outright damage to existing proofs of this title. At times, this task is quite monumental, requiring an almost total "rebuilding" of some pages from digital proofs of multiple copies. Despite this, imperfections still sometimes exist in the final proof and may detract from the visual appearance of the text.

I hope you enjoy reading this book as much as I enjoyed making it available to readers again.

Roger Chambers

CONTENTS

	PAGE
Preface	i
The Home Radio: How to make and use it	1
Principles of wireless telephony	6
Receiving sets	15
Reading diagrams	23
Tools and supplies required	25
Aerials and how to install them	30
Air-gaps and lightning switches	36
Counterpoise	40
Condensers	43
Transformers	48
Ammeters	51
Inductances and tuning-coils	53
Vario-couplers and variometers	58
A simple crystal detector receiving set	64
Crystal detectors and how to make one	67
Loose-coupled coils	68
Amplifiers	74
A simple vacuum tube receiving set	77
Regenerative receiving set	80
Transmission or sending	84
The simplest sending set	89
Another simple transmission set	91
An efficient 5-watt transmitter	94
Useful things to remember	98

PREFACE

THIS book is not intended as a treatise on radio telephony nor has any attempt been made to enter deeply into an explanation or discussion of the scientific phase of radio transmission. It is intended and designed particularly for the use of amateurs, boys, and those who wish to know how to make, use or adjust wireless telephone instruments and who are not interested in, or do not care to learn, the technicalities of practical electricity as applied to wireless work. A deep or thorough knowledge of the subject is not at all essential in order to secure all the pleasure and benefit to be had from the use of radio telephones, and unless the reader is well versed in a knowledge of the basic principles of electrical mechanics and phenomena, any attempt to explain the whys and wherefores of the subject would merely lead to confusion. Indeed, very few of the people who are using wire-

PREFACE

less telephones at the present time have any idea as to why the "wheels go round" and I am not at all sure that a complete ignorance of the inner workings of the apparatus is not a distinct advantage. As soon as a person once learns something of the mysterious force which renders radio telephony possible there is an overwhelming desire to experiment, and unless one is using radio for the sole purpose of experiment, or else has a good knowledge of the subject, such attempts usually result in complete failure of the instruments. Hence, the author has purposely avoided all technical terms and dissertations on the functions and reasons for various appliances and has aimed merely to make his directions and explanations as short, plain, and simple as possible, with purely diagrammatic figures to illustrate the subjects treated.

The subject of wireless telephony is a very large one and volumes might be and have been written filled with involved and technical explanations, formulae, etc. These are all very well for the professional or the expert, but the present work is not intended for these, but for the amateur.

PREFACE

Today, thousands of men, women and boys, as well as girls, are using wireless telephone receiving sets successfully while thousands more are anxious to do so. Far too many also, are using transmission or sending apparatus and a very large percentage of all these are interested in improving the sets they have, in installing more efficient sets, in making sets of their own or in similar matters, and it is to fill this want that this book has been written.

Wireless telephony has spread like an epidemic throughout the land and has come upon us, as an every-day diversion, without warning, but there are still vast improvements to be made before it will become as widespread and as important as telephoning over wires and it is highly probable that many of the greatest inventions and improvements of the future will come from amateurs who, by experimenting, chance upon undreamed of things. There are limits to wireless telephony, however, as we know it at present, and users should not expect too much from it. It is hopeless for a purchaser of a cheap set to get the results that might be reasonably expected from an expensive

PREFACE

set, and unfortunately, too many manufacturers, knowing the ignorance of most people on the subject, have taken advantage of this and of the tremendous popularity of radio telephony, to make misleading or false claims for the ready-made sets. It is hoped that this book may serve as something of a guide to those contemplating the purchase or use of instruments, for by learning what types of sets and of appliances are most efficient and the advantages and shortcomings of various types, they will be saved a great deal of disappointment and not a little time and money. Many people are under the impression that wireless telephony is as highly perfected, as useful and as simple as wire telephony, but this is not the case. Unless one uses an expensive set with amplifiers, the sounds from any but nearby stations are faint; they are often interrupted or broken by the interference of other signals, static, etc., and its use is entirely confined to listening-in to broadcasted messages or to communicating with amateurs or friends. The science has not yet reached that stage where one may call up and talk with other people,

PREFACE

as over an ordinary telephone, and it will doubtless be a long time before that is possible. Indeed, it is very doubtful if radio telephony will ever supersede wire telephony any more than radio telegraphy has superseded cables or wire telegraphy. There are many reasons for this, among which is the difficulty in preventing anyone and everyone with receivers from listening in. It is exactly as if all telephones were on the same party line.

To be sure, various appliances have been devised to obviate this, such as coding machines, machines which leave out certain sounds so that messages are unintelligible save to those provided with a synchronized instrument to reproduce the sounds properly, etc. But despite these, there will always be the fact that the air is filled with messages which anyone may pick up, just as any wireless telegraph operator may pick up messages and which has never been obviated, despite the time which has elapsed since wireless became a commercial factor.

But discussions of such matters really belong to a volume of a very different type from this work, which is not in any sense

PREFACE

a how or why it works book, but rather, a how to do and make it book. If it serves to enlighten its readers on many points not clear to them, if it enables them to make and install outfits which give pleasure or satisfaction, or if it serves to increase an interest in wireless telephony among beginners and amateurs, its purpose will be amply fulfilled.

The author wishes to express his appreciation to the Radio Editor of the New York Globe, to Wireless Age and other publications, and to Geo. Louis Gates, Floyd Rittman, Geo. A. Grogan, F. C. Greenwald and many others for valuable data, ideas, hints and devices which have been of great help in the preparation of this book.

THE HOME RADIO:
HOW TO MAKE AND USE IT

THE HOME RADIO:
HOW TO MAKE AND USE IT

WE usually think of the Great War as a disaster and as having done an incalculable amount of harm to the world and its people, but it is very doubtful if the war did as much harm as good. Within the few years it lasted it resulted in immense benefits to mankind in the tremendous discoveries, improvements and advancement of medicine, surgery, chemistry, aeronautics, mechanics, engineering, metallurgy, and wireless communication and while the toll of life, the destruction of property, the devastation of lands, the loss of art and the bankruptcy of nations are all temporary and will soon be replaced and forgotten, the scientific progress and discoveries, which were a direct result of the war, will endure forever and will continue to benefit mankind.

Of all the arts and sciences which

received an unprecedented impetus by the war, none is of greater interest or has a more far-reaching effect to the average person than radio telephony. Before the war, wireless telegraphy was well established, universally used and a fairly exact and well-understood science; but wireless telephony was scarcely more than a dream, —a visionary, uncertain thing; complicated, little understood and literally in its infancy.

But today, so incredibly rapid have been the strides made in the development of this science, that wireless telephony is an every-day affair; a simple, easily understood thing,—far simpler than telegraphy —and in constant use, not only commercially, but by countless thousands of amateurs. In a way, however, the war merely launched practical radio telephony on its career and far greater progress has been made in commercializing, simplifying and cheapening it within the past four months, —October to February, 1921-22—than within the previous six years. Indeed, so rapid has been its advancement, that it seems to have come upon us almost overnight, and within a few short weeks it has leaped from an obscure, scientific curiosity to an almost universally-used means of entertainment and profit.

Today, thousands of mere boys are using wireless telephones,—many of the instruments made by the boys themselves,—and in department stores, electrical supply stores and elsewhere, sets for receiving may be purchased for a few dollars. From various stations, music, crop reports, market reports, weather reports, speeches, songs, operas, plays, stories, official time, racing, and base-ball returns; shipping news and countless other interesting matters are sent broadcast through the air, free to anyone who possesses a wireless telephone receiving set. Thus, the farmer, miles from the nearest town, the sailor at sea, passengers on ships, guests in hotels, crowds about bulletin boards and people in their own homes can listen to the voices of famous men, the music of bands and orchestras, the singing of famous operatic stars, the dialogues of plays and countless other things, from far-distant points and as clearly and plainly as though no space intervened.

THE HOME RADIO

Truly, radio telephony is the great modern miracle; a dream more fantastic and fairy-like than the Arabian Nights; a more marvelous actuality than the fabulous lamp of Aladdin or the flying carpet, and, best of all, it is within the reach of everyone, while the "music in the air" is free to all who care to listen in.

It is certainly a strange, almost incredible, thing to think that the air about us, even within our dwellings, is constantly filled with sounds, voices, music, messages and songs which are as inaudible as they are invisible, but which may be caught and delivered to our ears by means of a few wires and batteries and a few appliances so simple that even a child may use them.

And the limit is far from being reached. Within a few years, or even months, the range of wireless telephony will be increased by hundreds or even thousands of miles, antennae or aerials will be entirely done away with and the instruments for sending and receiving will be so improved, simplified and reduced that one may carry them in one's pocket, for while radio telephony has already become

HOW TO MAKE AND USE IT

highly perfected, widely used and absolutely practical, yet it is still hardly out of its infancy and no man may prophesy what its future may be.

PRINCIPLES OF WIRELESS TELEPHONY

BEFORE attempting to explain the functions and the principles of the radio telephones, or describing how to make, use and operate them, it is necessary to understand something of the underlying principles and fundamental laws of wireless transmission.

It is not, however, necessary, and in fact it is impossible in a book of this scope, to enter into a long discussion or dissertation on the theories and principles of electricity or physics which enter into the subject, but merely to illustrate and make clear a few important and salient laws, causes and results which make the transmission of sounds possible without the use of wires between the sending and receiving instruments.

The first and most important principle of all radio transmission is the fact that all our atmosphere is constantly disturbed by vibrations or oscillations or, as we may call them for the sake of simplicity, waves. We are accustomed to think of the atmosphere about us as a more or less uniform substance which we call air, but in reality, the air or atmosphere, space, and in fact all solids as well, are pervaded by an invisible, odorless and almost weightless but exceedingly elastic substance known as *ether* or *luminiferous ether*. It is by means of this ether that all heat, light, electricity, etc., are transmitted, in the form of waves or vibrations. Light and heat waves have been known to science for a long time, but it is only within comparatively recent years that man has learned that electric or electromagnetic waves also travel through the ether and it was through this discovery, and by means of these waves, that wireless telegraphy and telephony became possible. In a way, the electromagnetic waves are very similar to the waves or ripples formed by dropping a stone in a calm pool of water for, just as the water waves travel in gradually widening circles from the splash, so the electric waves, started by the spark of a wireless transmitting or sending instrument, spread in ever-widening circles through the ether. Moreover, just as the waves in the pool are short and clear near the spot where the stone is thrown and gradually become longer and less noticeable and farther apart as they flow from the splash, so wireless waves are clearer and sharper near the instrument and decrease in clearness and size as they get farther and farther away and, to draw still another comparison, just as the shore of the pool or any object in the water interrupts or breaks the waves, so a wireless receiving instrument will interrupt or receive the electrical waves of the ether. Indeed, just as the waves or vibrations set up by the wireless spark are electromagnetic waves and the energy that starts them is electrical energy, so the waves in the pool are started by muscular or mechanical energy. You can readily understand how it would be possible to communicate by means of such liquid waves, for, if a person at a distance should toss stones into the water at stated intervals, a person watching the shore, and noticing the intervals of the waves, could understand signals which had been prearranged. This,

HOW TO MAKE AND USE IT

of course, would be a very crude and uncertain method of communication; but if you could devise some instrument to count and measure the waves and could devise means for creating waves of definite sizes and numbers, a practical means of communication could be established. It is just this which occurs in the transmission and reception of electromagnetic or wireless waves, for, by breaking or interrupting the waves sent out by the spark they are transformed into long and short sections which correspond to dots and dashes as used in wireless telegraphy. These waves set in motion by the sending apparatus, possess the property of starting oscillations in any conductor which they strike, and if they strike the aerial or antennae of a wireless station they start oscillations in the wires, but so faint and weak that they would not be detected unless sensitive instruments were provided to magnify and catch them. Such instruments are known as *detectors* and consist of various substances or devices which are adjustable and from which wires lead to a telephone receiver. The high frequency oscillations of the ether, which are known as *alternating* currents, as they flow back and forth, are cut off by the magnets in the receiver, while the detector is designed to allow the oscillations to pass through in one direction but will not allow them to return, and thus it acts as a check-valve in a water pipe and the alternating, back-and-forth currents are transformed to impulses going in one direction only and known as direct currents. These will flow through a telephone receiver and cause the diaphragm to vibrate and thus the ear detects the interrupted buzzing sounds which indicate the dots and dashes of the code. Another very important part of the mechanism of the wireless apparatus is the tuner, for without this the various vibrations of the ether sent out from numberless stations would come as a hodge-podge of meaningless sounds to the receiver. But by arranging the receiving instrument so it may be adjusted to receive or pick up only those waves of a certain length, all other vibrations and the messages they carry being eliminated or cut out. So, the wireless operator who is receiving messages, may adjust his instrument back and forth until he picks up any message which may be passing through the ether at the time.

In many ways, all these principles and laws of wireless telegraphy apply equally well to radio telephony, but, in other ways, there are great differences between the two. In the first place, the vibrations or waves sent out by the wireless telegraph transmitter are interrupted as they are produced by the vibrator of a spark coil and while the flow of the oscillations or waves may be so rapid that they appear as a steady stream, yet there is really a distinct pause after each one. So, if a telephone transmitter and receiver were attached to the ordinary wireless instruments used in telegraphy, the sounds or vibrations of the voice would vary the path of the electrical oscillations and the same variations would be produced on the receiver. But, owing to the fact that the waves are interrupted and are not continuous, the words or sounds would be cut up into sections which would be meaningless, although such sounds as music, bells, whistling, etc., might be recognizable. In fact, sounds of this sort frequently *have* been heard over ordinary wireless telegraph instruments. Hence it is easy to see that the only reason why all sounds cannot be carried through space by ordinary wireless telegraph apparatus is because of the interrupted or broken waves, whereas, if these waves were continuous or were so incredibly rapid as to appear continuous, the sounds they transmitted could easily be heard and understood. Therefore, the real fundamental key to successful wireless telephony lies in producing what are known as *continuous waves* and it is to the perfection and control of such waves that radio telephony owes much of its rapid advancement.

The accompanying figures *Nos. 1, 2, 3, 4, 5, 6* illustrate just how the interrupted telegraphy currents and the continuous waves carry sounds. Thus, *1*, represents the variations in vibrations in a certain word. *2*, the intermittent oscillations of the wireless telegraph sender and *3*, the way the word would be broken and interrupted by being transmitted by means of such interrupted waves. *Fig. 4*, on the other hand, shows the continuous waves of the wireless telephone transmitter. *5,* the sound waves of a word and *6,* the way the continuous waves, interrupted by the words waves, would appear. By studying these diagrams you can easily see the difference between the word broken up as in *Fig. 3* and flowing smoothly as in *Fig. 6.* It must be borne in mind, however, that whereas the sounds of the voice, of music, etc., cannot be satisfactorily sent or received by wireless telegraph instruments, yet wireless telegrams *can* be sent to perfection, and can be perfectly received, over wireless telephone instruments. Indeed, the very best transmitting radiophone sets serve the best for sending telegraphic messages, the only difference being that for the former the continuous high frequency waves are used, whereas, when sending code messages by telegraph, the waves are broken or "chopped" by suitable instruments and a key which opens and closes the circuit.

In using wireless telephony, just as in wireless telegraphy, there must be two separate units known as transmitters and receivers and a transmitter can be used only for sending and a receiver for receiving. As the sending or transmitting apparatus of the wireless telephone is far more complicated than the receiving instruments, as the greatest interest in wireless telephony lies in receiving the messages, songs, etc., sent broadcast from large sending stations, and as the receivers are very simple and easy to understand or to construct, we will put the cart before the horse, so to speak, and consider the receiving end of radio telephony before we take up the sending end.

RECEIVING SETS

BROADLY speaking, the receiving set consists of the *antennae* or *aerial*, the *tuner*, the *detector* and the *receiver*, but aside from the aerial, each part of the instrument is made up of several other units and appliances, each devised and used for a definite purpose.

The aerial, which is a wire designed to interrupt or catch a portion of the continuous waves (always referred to in wireless telephone parlance as C. W.) consists of a single wire, for unlike wireless telegraphy, a number of strands or wires is of no advantage in receiving, and still more remarkable, it makes no difference whether the wire be bare or insulated, for the C. W. used in radio telegraphy penetrates solids of every kind. Indeed, a wire stretched around a room or through a hallway indoors, or even a metal bedstead or bed spring, may serve as an aerial for receiving wireless telephone messages, although far better results are

THE HOME RADIO

secured by properly installed aerials out of doors. And here it may be wise to impress all users of receiving sets with the fact that the longer the aerial and the higher above the ground the better will be the results obtained, although an aerial 150 feet long and well above other large buildings will serve every purpose. Moreover, it makes no difference whether the aerial is horizontal, vertical or at an angle, provided it is thoroughly insulated from all surroundings, and very good results have been obtained by aerials run vertically up an air shaft or along the side of a building. So too, the *lead-in,* or wire connecting the aerial with the receiving instrument, serves as an aerial itself and therefore a long *lead-in* with a short aerial will serve almost as well as a *long aerial* and *short lead-in,* which is a tremendous advantage to dwellers in hotels, apartment houses, etc., where it is very difficult or impossible to install a long and lofty aerial on the roof. But before going into details and describing the installation of aerials, let us consider the rest of the receiving equipment and thoroughly understand its principles.

HOW TO MAKE AND USE IT

The *detector*, without which it would be impossible to register or detect the minute currents or waves which pass through the aerial, is a very important part of the mechanism. There are two types of detectors in use, the first known as the *crystal detector* and the other as the *vacuum tube*. In the former, a crystal of some mineral—preferably galena—is used, while in the latter, a form of incandescent lamp with especially prepared filament is employed. Of the two, the former is the cheaper and is used on most of the cheap, ready-made sets; but it has limitations and is not nearly as satisfactory in many ways as the vacuum tube, as will be explained later. The third unit or *tuner* is the means by which the entire apparatus is made to pick up the sounds from some station or elsewhere and by means of which other sounds are shut out, for only by means of the tuner can the receiving set be placed in *synchronism*, or "tune" with the waves carrying the sounds you wish to hear. The last unit or *receiver* is merely a telephone receiver made for the purpose and which, on a small set, is worn over the ears exactly as in receiving wireless telegraph messages.

Fig. 7 illustrates in diagrammatic form these units with their relative positions in the set, but there are several other accessories which are essential to good results.

If the user of a receiving set purchases his outfit ready-made, there is no necessity for knowing how it is constructed or the functions of its various parts, for all that is necessary is to install it, adjust the detector and tuner and listen. These ready-made sets, moreover, are accompanied by full directions for setting up and using, so that advice on these points is scarcely necessary. However, many amateurs have difficulties or expect too much for their money and a few hints and cautions may not be amiss. If the set is a *crystal detector* set, the *detector* will be found to consist of a bit of crystal held in a metal stand or cup with a fine wire and screw making a contact with it, *Fig. 44,* although in some types two crystals are used to make contact with each other. To adjust the *crystal detector* the movable contact must be shifted about over the crystal until a sensitive spot is found, which is easily determined by the strength or loudness of the sounds carried to your ears by the receiver. The *tuner* is very easy to adjust as it is provided with a knob or handle, and by turning this slightly until the sounds come clear and uninterrupted, all other sounds or waves are cut out. Then, by marking the handle, you can always pick up the same station by placing the tuner at the same point.

These *crystal detector* sets, however, have several *disadvantages*. In the first place, their range is limited to a comparatively short distance, depending largely upon the length and height of the aerial. In the second place, it is practically impossible to tune out interference from nearby stations. In the third place, the least jar or vibration will disturb the point of contact on the crystal and a new adjustment will have to be made; while in the fourth place, such crystal detector sets can only be used with a head telephone receiver and cannot be *amplified* or built up so as to produce loud sounds audible all over a room or to persons not using a receiver. It should also be borne in mind that a *buzzer*—which may be bought ready-made or may be made by removing the gong from an electric bell—is a very useful instrument for adjusting a crystal detector. The buzzer may be operated by a single dry battery cell and as the spark at the contact maker of the bell or buzzer sends out electrical waves which will affect the detector, it is very easy to select the most sensitive spot on the crystal by this means.

Far better than these crystal detector sets, are the so-called *vacuum tube* sets in which a specially constructed electric light with a metal sheath and a coil of wire around the filament takes the place of the detector of crystal. To use these sets, it is only necessary to adjust the tuner and the brightness of the filament, but care should be used in doing the latter as they are very delicate affairs and are expensive and a filament may soon be ruined by improper adjustment or use. If the filaments burn too brightly you are shortening the life of the tube without gaining anything and tubes are the most expensive part of the outfits.

HOW TO MAKE AND USE IT

The way to do is to turn the filament-adjusting knob very gently until you receive the maximum and clearest messages and do *not turn the knob any further*. If you do so, you will not increase the tone, but will cause the machine to squeal and howl, in addition to burning out the filament. Very often, a beginner will turn the knob too far and then, when no messages come in, he gets excited and turns first one thing and then another with no result, all because he has given such a shock to the tube that it is paralyzed, for the vacuum tube detector is one of the most delicate instruments ever invented and must be handled accordingly. Luckily, the tubes have great recuperative power, and if left alone for half a minute or so, they will come back in their normal condition, when proper adjustments may be made.

As soon as you have adjusted the tuner and filament-adjusting knobs so you receive the messages clearly and distinctly, mark each knob so that you can pick up the same station next time and thus, by marking the point for every station within receiving distance, you can readily adjust your instrument for whichever one you wish to pick up. Of course you may have to do some fine minor adjusting after the station is caught, for the weather and atmospheric conditions vary and consequently tuning or adjusting is always necessary; but by having the knobs marked you will save a vast amount of trouble and time.

READING DIAGRAMS

MOST people who have not made a study of radio telegraphy or other branches of electricity are puzzled when they look at the diagrams for wiring that are supplied in many books and magazines or catalogues. These seldom have the various appliances or accessories marked by name or letter and, to the uninitiated, they are practically meaningless. It is very easy to understand these, however, once you have learned what the various symbols mean, and everyone interested in radio telephony should learn them. In the accompanying cut, *Fig. 8,* the commoner symbols used in diagrams of wireless apparatus are shown and anyone can learn and memorize these in a short time.

In nearly all diagrams of wiring and setting up radio telephone instruments, the wires are drawn parallel with one another and with turns at right angles. This adds greatly to the appearance of the diagrams, but in actual practice it is a great advantage *not* to run the wires parallel or with the turns at right angles. For this reason, in the majority of the figures I have given, the wires are shown at angles.

TOOLS AND SUPPLIES REQUIRED

TO give an entire or complete list of the various tools and supplies required for making, setting up and using wireless telephones is practically impossible. In the first place, some people can work advantageously with fewer tools than others; some people are naturally "handy" or inventive and can find uses for odds and ends which would appear worthless to others; some people must economize on tools and supplies, others can spend an unlimited amount, while still others prefer to purchase most of their appliances ready-made and merely put them together or set them up.

For these reasons, the tools and supplies listed below, are only those which will prove most necessary and as their quality, size and number will depend largely upon the work to be done and one's pocketbook, no prices or estimates of their cost have been given.

TOOLS

One large screwdriver.

One brad-awl set of awls, screwdrivers, etc., or small and medium sized screwdrivers.

One gimlet.

One hack saw frame and saws.

Panel or cross-cut saw.

Mitre-saw and mitre-box.

Chisels and gouges.

Three-cornered file.

Round or rat-tail file.

Flat file.

Sandpaper.

Smoothing or block-plane.

Small bench (iron) vise.

Bit-stock with bits and augers.

Breast or hand, geared drill with twist drills.

Flat-nosed pliers.

Round-nosed pliers.

Cutting-pliers (flat-nosed and cutting pliers combined may be used).

Soldering iron, solder and flux.

Tack hammer.

Claw hammer.

Carpenter's square.

HOW TO MAKE AND USE IT

Tape, yard stick or rule.
Set of small screw-taps and dies.
Compasses or dividers.

SUPPLIES

Wire nails.
Wood screws (flat head), assorted steel or brass.
Wood screws (round head), assorted brass.
Washers for round-head screws.
Small brass bolts and nuts, assorted.
Emery paper.
Wire of various sizes (see directions), copper, plain or bare.
Same insulated (see directions).
Stiff cardboard.
Paraffine wax.
Good glue.
Sealing wax.
White shellac.
Fibre board or bakelite.
Hard rubber knobs.
Flexible insulated wire cord.
Porcelain insulators.
Tin foil.
Binding posts.
Terminals.

Varnished cambric tubing.
Strong twine or string.
Sheet brass or brass strips (see directions).
Sheet copper.
Adhesive tape.

The last is one of the most useful articles one can have. It is useful in wrapping joints of wires; in covering wires as an insulator; in attaching wires where they cannot be soldered; in making temporary joints or connections; in covering coils or holding the wires on coils in place; in holding parts of cases or boxes together while they are being glued or nailed; to cover a cut in your finger, as well as for a thousand and one other purposes. But do *NOT* use the cheap, weak grades of tape sold in ten-cent stores and in many bicycle and automobile accessory dealers'. Use a good, strong, rubber-covered tape such as *Tirro,* for while it costs more it is worth many times as much more. The cheap tapes dry up, lose their stickiness upon exposure to air; the thin rubber, if any, soon disappears and leaves only the fabric which is not an insulator and they have no tensile strength, whereas the high-grade tapes are exceedingly strong, they are coated heavily with rubber, they never dry up and they retain their tenacity for a long time.

Varnished cambric tubing, known also as "Spaghetti", is the best material for covering joints in wires and should be used wherever possible.

Finally, let me advise you *never* to throw away anything which you have on hand in the way of electrical supplies, wires, screws, nails, etc. One never knows when such things may come in handy and may be put to some good and useful purpose, thus saving time and money.

AERIALS AND HOW TO INSTALL THEM

ONE of the greatest advantages of wireless telephone receivers is that an elaborate or expensive aerial is not required. Although good sets with vacuum bulb detectors may be used with an indoor aerial, or even with a bedstead or wire springs as an aerial, yet an outside aerial will always give better results. As I have already mentioned, a single wire will do as well as several, the main thing being to get the aerial long and high in order to catch waves which are not interrupted or interfered with by surrounding buildings, steel bridges, electric wires and similar objects. Next, or rather most important, is to have the aerial and *lead-in* thoroughly insulated from all surrounding objects, for even wood, when damp, is an excellent conductor. The best material for an amateur aerial for receiving is a stranded phosphor bronze or copper wire, about No. 14, although solid copper wire, copper-cov-

HOW TO MAKE AND USE IT

ered steel wire or even insulated copper wire will serve every purpose. For insulators, use porcelain cleats. These may be used both where the lead-in is attached to walls or other objects, and where the aerial wire is attached to the supports or guys. The accompanying figures, *No. 9 and No. 10,* illustrate aerials installed, the first showing the wire attached to a chimney or similar structure and to a wall; the other, an aerial which is designed for a tin or slate roof and which obviates making holes for attachment. Where the lead-in wire enters the building it should be of *rubber insulated wire* and may be brought in at the corner of a window, either by cutting a small groove or by jamming the window down until the wire flattens and is buried partly in the wood. All joints in the aerial and lead-in should be *scraped bright, tightly twisted* and *soldered,* finally being wrapped with insulating or adhesive tape or covered with "spaghetti" tubing. For the best results, be sure to run your lead-in from the end of aerial towards the station which you most frequently wish to hear or towards the most distant station which you desire to pick up. Very often, this will make a vast difference in results, especially with a small receiving set. If there are several sending stations at various points from your set, it is often a very good plan to run several aerial wires at right angles or radiating as shown in *Figs. 11 and 12*, connecting them together and running the lead-in from the point where all join, as shown. Sometimes this principle may be reversed and several lead-ins may be carried from the outer ends of the radiating aerials and joined to form a single lead-in and will bring even better results, *Figs. 13 and 14*. These several lead-ins may be connected by means of an *anchor-ring, Fig. 15,* to equalize the waves or currents, or they may be provided with multiple-point switches as shown in *Figs. 13 and 14*. This switch arrangement has the great advantage that you can largely cut out stations you do not wish to hear by using the lead-in towards the station you desire to hear. This will result in the others being fainter or weaker in comparison and they can therefore be more easily tuned out by your instruments.

HOW TO MAKE AND USE IT

Aerials are most peculiar affairs and a little experimenting will enable you to determine the best size, height and type to use. It is well known that wireless waves are directive, or in other words, that they travel more strongly in one direction away from the sending aerial than in others and while this has been largely obviated in up-to-date stations, yet the ordinary receiving aerial is directive and will get stronger signals if the lead-in is towards the sending station, or is pointed towards it, so to speak. I know of several cases where amateurs failed utterly to hear voices, music, or even telegraphic spark signals from some station and yet, merely by altering the direction of their aerial or the position of the lead-in they could hear everything perfectly. So you see a great deal may depend upon the simple aerial, even if it consists of only a single wire. On the other hand, many amateurs have obtained splendid results with a wire run around the walls of a room near the ceiling; a wire run through a hallway; a wire dropped down an air-shaft or elevator-shaft, or even from an iron bedstead or bed-spring. It all depends so much upon local and climatic conditions, surroundings and other conditions that no hard and fast rules can be made, but despite all this, nine times out of ten, a high aerial, well above surrounding buildings and from 100 to 150 feet long, will give the best results. But remember that if there are elevated tracks, steel bridges, trolley lines, electric wires or steel structures near, you should run your aerial at *RIGHT ANGLES TO THEM* in order to avoid failure through leakage or inductance.

You must also bear in mind that the "*ground*" is almost as important as the aerial, for without a good ground the set will not work. A water, steam, or gas pipe will usually make an excellent ground, but before using it be sure there is no insulated joint between the connection of your wires and the earth or that the pipe does not enter an earthen or tile pipe near the ground or in the cellar. In making the ground connection, scrape the pipe clean and bright and *solder* the wire to it. If this is not possible, wind the connection with tin-foil and fine wire and wrap it with adhesive tape. Where no pipe is available carry the ground wire to a sheet of copper, an old copper boiler or a copper tank or basin filled with charcoal and buried at least five feet under the surface of the earth. A lightning rod or fire escape will sometimes make a very good ground. But it is not so much *what you use for a ground as HOW GOOD THE CONNECTIONS ARE AND HOW WELL THE OBJECT IS GROUNDED. DO NOT* use an electric light or telephone, telegraph or door bell wire for a ground.

AIR-GAPS AND LIGHTNING SWITCHES

A GREAT many people are very much afraid of lightning following an aerial and injuring the premises, for they seem to think that the wires "attract" lightning, just as many people with intelligence and education still believe that steel knives or hardware or wire netting window screens "attract" the lightning. As a matter of fact, none of these things "attract" the lightning, but merely form a convenient conductor to enable the lightning to ground itself. Lightning-rods are designed for the same purpose and a properly installed aerial, instead of jeopardizing a building, is really an excellent safeguard and makes a splendid lightning rod. Lightning strikes a building or object when it is trying to find a way to the earth and if the object struck is a good conductor of sufficient capacity it does no damage. For this reason, houses covered with wire netting and climbing vines are far safer than those which are bare, and steel buildings, such as the New York skyscrapers, steel bridges, elevated structures, iron smokestacks and chimneys, iron steamships and railway tracks are seldom injured by lightning although frequently "struck;" the reason being that the electricity passes through them freely without encountering resistance. On the other hand, wooden buildings, trees and human beings are poor conductors and when dry are almost non-conductors of electricity, and when the lightning tries to follow such objects to ground, the resistance is so great that serious damage is done. It is exactly like forcing water through a pipe. If you have a powerful stream of water or a great volume of water and provide a pipe large enough for it to flow freely, the pipe will not be injured, even if it is very light and frail; whereas, if you attempted to force the same stream or same volume through a much smaller or clogged pipe, the pipe would be burst or the water would overflow and flood the surroundings. Statistics prove that as far as aerials are concerned there is no danger and records of fires or injuries from aerials during thunder storms are extremely rare. During an electrical storm the instruments cannot be used owing to the "static" or electricity in the air and the confusion of currents, waves and inductance, and by installing a *lightning-switch* or an *air-gap* there will be no danger to the premises. In fact, a properly installed aerial does not affect the rate of insurance and if installed in accordance with the regulations of the local fire department you may be sure there is not the least danger. The fire department records of New York City do not show a single instance of conflagrations started by aerials and lightning.

The simplest and best safeguard for receiving aerials is the air-gap shown in *Fig. 16*. This consists of two metal attachments separated by about one-eighth of an inch *A-B*, one of which (*A*) is attached to the lead-in wire (the wire to set being fastened to it also) while the other (*B*) is connected by a wire to the ground direct. This gap is mounted in much the same manner as a *lightning-switch*, *Fig. 17* (on a window sill or other convenient spot), in which *A* shows connections to aerial, *B* to receiver and *C* to ground connection. When the station is not in use, or during thunder storms, the handle *D* is thrown from *A* to *C*, this cutting off all connection between the lead-in wire and the instruments and connecting the aerial directly with the ground.

Aerials for sending or transmission stations are very different from those used for receiving only and should be of several wires. The most efficient is probably the *"cage"* type shown in *Fig. 18*, but any of the others illustrated in *Figs. 19, 20 and 21* will answer. These should, of course, be fully insulated and the various methods of doing this are well shown in the figures and require no explanation. To install sending stations a license is required, whereas to receive, no license is needed.

COUNTERPOISE

BEFORE leaving the subject of aerials it may be well to call attention to the device known as a *counterpoise* and which, for sending, is far superior to using a ground, while with small sets the advantages gained by a counterpoise in receiving do not pay for the trouble of installing the device. This is because the counterpoise, while adding to the sharpness of tuning with a receiving set and, therefore, aiding in cutting out interference, will also cut down the strength of the sounds received. Therefore, with a crystal set where amplification is not possible, the device is practically valueless, whereas, with a vacuum tube set with two or more steps of amplification, the counterpoise will prove a very distinct advantage. Many people consider this device a complicated and difficult affair, but in reality, it is as simple, if not simpler, than an aerial. A favorite form of counterpoise consists of several wires extending fanwise as shown in the figures, but a single wire will often give excellent results and the only way to determine the best number of wires to use is by experiment. Usually it is desirable to place the counterpoise below the aerial, but this is by no means essential as it may be run in the opposite direction from the aerial and still work exactly as well, for the device has little or no connection with the aerial. In fact, its action is more like that of a condenser, except that it increases radiated energy, whereas, a condenser has a very small amount of radiation. It must also be borne in mind that with a counterpoise *no ground wire* is required, the lead-in from the counterpoise being connected with the set at the spot where the ground wire is usually connected. In setting up a counterpoise it should be just as well and as thoroughly insulated as the aerial (*Fig. 22*), and the lead-in wire from it should be kept at some distance from the aerial lead-in to obviate losses by induction between the two. The most desirable place for a counterpoise is about three feet above the earth, but as this height is usually inconvenient, not only on account of it being an obstruction, but because it may be injured by people or animals or may be buried under snow in winter, it is better to raise it about six feet, or just high enough so people may pass beneath it. Stout posts with guy wires are the best supports, whereas, if the device is placed on the roof, the supports may be chimneys, walls, etc. If placed on a roof beneath an aerial leave all the space possible between the two, either by keeping the counterpoise low or raising the aerial. Where this is not convenient, the counterpoise may be run in another direction instead of being placed below the aerial wires.

CONDENSERS

THESE appliances are a most important part of a wireless set, as without them the oscillations, even if detected by the instruments, would be very weak and faint. They are divided broadly into two classes known as *Fixed Condensers* and *Variable Condensers,* the former being the simplest, and the latter the most efficient, for while a fixed condenser is always of one capacity and can only be increased or decreased by adding to or subtracting from the number of sheets, the variable type may be altered or adjusted at will by a knob or handle, thus tuning or adjusting the receiving circuit exactly as a tuning coil is adjusted, but much more delicately, as the adjustment of a tuning coil consists in shortening the length of coil by jumping connections from one turn of wire to another to alter wave lengths, whereas, the condenser adjustment is slow, even and gradual and alters capacity; but it must not be forgotten that for wireless telephony receiving, both a condenser and some sort of coil or similar device must be employed to get satisfactory results. The simplest form of fixed condenser consists of a number of alternating sheets of tin-foil and waxed paper or mica, the alternate sheets of foil being connected by wires which are in turn connected with the terminals where required.

To make a *fixed condenser* it is only necessary to lay sheets of tin-foil between sheets of waxed paper and connect them. A very efficient little fixed condenser, to be used by shunting across the receivers of a small set, or as a grid-condenser with a vacuum tube set, can be made as shown in *Fig. 23*. Have some smooth tin-foil, free from holes or tears, and cut two pieces about one inch in length and one-half an inch wide. Then, from thoroughly waxed paper—which can be purchased or can be made by soaking good quality writing paper in melted paraffine wax—cut three pieces 2½ inches in length and 2 inches wide. On one of these pieces place a piece of the foil; then cover this with a second strip of paper, place the other strip of foil over this and cover with the last strip of paper. Be very sure that the edges of the tin-foil are well within the margins of paper and are accurately in line or centred. In fact, before placing them, it is wise to draw a square the size of the foil on each piece of paper, spacing it equidistant from edges, and arrange the foil to fit this, *Fig. 24*. Next cut two pieces of light wire five or six inches long (flexible stranded wire is best), spread the strands at one end of each piece apart and place one of these frayed and spread ends on the lowest piece of foil between it and the bottom paper. Next, place the other, frayed and spread the same way, on the upper piece of foil at the opposite end and with a few drops of hot paraffine fix them in place on the edges of the paper. Then, roll the whole, being careful not to displace the foil (this may be secured to each piece of paper with a few drops of paraffine at the edges) and form a small cylinder *Fig. 25*. Wind the cylinder tightly with fine thread at each end, as shown, or wrap with adhesive tape and dip the whole into hot paraffine. In using this condenser with a crystal set it is only necessary to shunt, or connect it, across your phone receiver wires, but if using it in a vacuum tube set you must use a *grid-leak* shunted across it. This is merely a piece of cardboard placed between two binding-posts or terminals and with several soft lead-pencil lines drawn across it from post to post. In order to be sure that a good connection is made, draw pencil marks about the holes where posts are to be inserted. The distance between posts should be not over five-eighths of an inch. Sometimes drawing ink (made of carbon, for writing ink will not serve) is used in place of lead pencil, but the latter has the advantage that the lines can be varied or adjusted to give best results by means of an eraser.

VARIABLE CONDENSERS are much more difficult to make and while any ingenious boy *can* make them, it is usually cheaper to purchase them ready made. There are two common forms, one known as the *sliding plate, Fig. 26,* the other as the *rotary, Fig. 27.* The former consists of a number of metal plates, which slide back and forth in a frame, case or box provided with grooves and

HOW TO MAKE AND USE IT

fixed plates. The rotary type consists of a number of semicircular plates of metal so arranged as to rotate or swing past a series of fixed discs. In each form, the air spaces between the plates correspond to the waxed paper between the strips of foil on the fixed condensers. By means of either of these two variable forms, fine adjustment of capacity may be obtained. Many people cannot understand the function of a condenser, but, broadly speaking, it is to store up electrical energy and then suddenly release it, as the current passing through is interrupted, varied or broken, or, in other words, to increase the oscillations.

Although most small sets will operate with a fixed condenser and a variable one is not necessary, yet the variable type will always improve the receiver and will permit much finer tuning than a tuning coil of any type by itself.

TRANSFORMERS

THESE are instruments designed to transform or change one kind of electrical current to another such as alternating current to a direct current, and are very useful and essential devices in radio telephony. There are many kinds of transformers, but all are built, or rather based, upon the same principles, which is that of inductance, or the formation of a current in a coil of wire by the passage of another current through another coil near it. As induced currents are only produced when the magnetic field is changing, the current induced by a transformer can only be secured by means of some mechanical device or by an alternating current. When the former is used the transformer becomes a *spark-coil* or *induction-coil* (see coils) and the means by which the primary current is alternately broken or interrupted is the buzzer or contact at the end of the iron core of the coil. But if an alternating current is run through the primary wires of a transformer no interrupter is required, as the magnetic field changes each time the current rises and falls. There are two general types of transformers in use; one known as an *"open-circuit transformer"* which is exactly like an ordinary sparking coil and consists of an iron core covered with two windings of wire known as the primary and secondary, *Fig. 28, A*. Very often, where such a transformer can be used, an ordinary spark-coil with the contact-breaker screwed down answers every purpose. The other type is known as the *"closed-core transformer"* and consists of a number of iron plates or laminations in the form of a hollow square and which are wound on one side for the primary and on the opposite side for the secondary. *Fig. 28, B*. Although either one of these types may be made at home, yet it is not advisable to attempt it. In the first place, several thousand turns of secondary wire are required and it is a tedious and difficult matter to wind these on evenly and well. Moreover, the number of turns of primary and secondary wire must be very carefully proportioned and must be worked out on mathematical lines in order to secure the proper reactance or the tendency to resist the flow of the alternating current and induce the direct current. Finally, transformers are not expensive and the cheapest are far more efficient and are better made than anything you can make yourself.

AMMETERS

AMMETERS are instruments designed to measure the flow of electricity through the wires and are often very essential parts of a radio outfit. They consist, as far as exterior appearances go, of a dial marked with figures and a hand or needle. Although they are not high priced and it is not advisable to try to make them, still there is nothing mysterious or complicated about them and the amateur, who likes to experiment with home-made instruments, can readily construct an ammeter which will work and is fairly reliable. This instrument is known as a *"hot-wire ammeter"* (*Fig. 29*), and consists of a fine platinum wire A, secured between two fixed supports B, B, a thread C, fastened to the centre of the wire and passed around a spool or spindle D, a spring attached to the end of the thread E, and a pointer or hand fastened immovably to the top of the spindle F. The electrical connections are made at *B, B,* and as soon as a current passes through the platinum wire, *A,* the wire becomes heated and expands, thus allowing the thread to slacken. The slack is instantly taken up by the spring *E,* thus revolving the spindle and swinging the needle to one side. The greater the current the more the pointer swings and so, by arranging a dial with marks under the needle and testing the device with currents of known force, a fairly accurate instrument can be made. Needless to say, the parts must be small and neatly and accurately made and the spring must be adjusted to merely hold the thread tightly without pulling or bending the wire appreciably. Also, the length of the wire is a great factor for the longer the wire the greater will be the amount it expands, and consequently, the greater the movement of the needle; but, on the other hand, it will be more difficult to adjust a long wire to remain tight than a shorter one and the only way to determine the proportion of the various parts is to experiment.

INDUCTANCES AND TUNING COILS

IN order to receive and hear sounds sent from transmitting stations by radiophones clearly and without interference or confusion, a device of some sort is required which will cut out all waves save those desired. This is known as "tuning" and the instruments or appliances used to accomplish it are called "tuners." There are now a great many different devices for tuning, such as tuning-coils, loose couplers, vario-couplers, variometers, variable condensers, etc. Of these, all but the variable condensers (which see) are coils of various types, the simplest, but by no means the most efficient, being the simple tuning-coils. These consist of a coil of bare wire wound about a core or cylinder of wood, fibre or pasteboard and provided with sliding contacts as shown in *Fig. 30*, in which *A* is the coil, *B*, the slide rod and *C*, the slider.

By moving the slider from coil to

coil of the wire the wave length of the receiving instruments may be adjusted to catch the desired sounds of that wave length. Such coils are very easily made by winding a pasteboard tube—which should be soaked in melted paraffine to render it waterproof—with bare copper wire about No. 18, making about 40 turns and leaving a space of about 1-16 inch between the turns. If two or more sliders or contacts are arranged still finer adjustment will be attainable while, by providing rotary switches with five contacts as shown in *Fig. 31,* still better results will be secured. Although, as stated, these coils are easily made, yet they are so cheap that many prefer to purchase them ready-made rather than bother making them. Simple coils or inductances, however, are even simpler, as they consist merely of a few turns of insulated copper wire wound on a pasteboard tube, the number of turns depending upon the wave lengths to be received. In some sets there is but one coil or helix *Fig. 32,* while in other sets there are two, a primary and secondary *Fig. 33,* and as a rule the coil should be *tapped* and the circuit con-

HOW TO MAKE AND USE IT

nected at the *tap-off* **Fig. 34**. This is best done by taking a loop or twist in the wire at the desired point and then continuing winding as **Fig. 35**. Of course, in making the connection at this tap-off the wire should be scraped free of insulation to make the joint, after which it should be wrapped with adhesive tape. In making these simple inductance coils it is best to put on more turns of wire than you think is actually required, as it is far easier to remove one turn at a time, until the desired wave length is obtained, than it is to add turns after the instruments are set up. Similar simple coils are used in many parts of receiving sets, such as the radio-choke in **Fig. 59, L,** etc.

By making several tap-offs and then leading them to the various contacts of a multiple-point switch, **Fig. 36, C,** excellent results may be obtained especially with the smaller crystal sets with a single simple coil. Another way by which waves of varying lengths may be received by means of simple coils without tuning devices, is to have several coils of various sizes so arranged that they may be connected or disconnected with your set at will. This may be done, either by means of plugs and sockets as in *Fig. 36, A,* or by switches with several contacts as shown in *Fig. 36, B.* Still finer adjustment may be obtained by providing each coil with a slider or similar tuning device. This will give a wide range of wave lengths and will obviate all need of taking turns off the coils and as such coils are very easy to make you can have as many as you desire of different sizes.

Another very different type of *inductance-coil* consists of two windings, known as *primary* and *secondary*, and are similar to the ordinary sparking coils used in older type automobiles, in power boats, etc. These are known also as *transformers* (which see) and while they *can* be made at home yet it is a tedious and difficult job to wind on the hundreds of turns of wire properly and as such coils are inexpensive it is never advisable to attempt it, unless you wish to make everything yourself, just for the practice and fun of it. Coils of this type may be purchased which are made especially for radio use, but an ordinary *spark-coil* with the *contact-breaker* screwed or fastened

HOW TO MAKE AND USE IT

down will answer every purpose. The same type of coil, using only the secondary winding, may be successfully employed as a *choke-coil*, as shown in *Fig. 60, N.*

VARIO-COUPLERS AND VARIOMETERS

VARIO-COUPLERS may also be home made if desired, but they are seldom very efficient and as they are very low in price it is scarcely worth while to attempt their manufacture. These consist of two coils, one rotating within the other. If desired to make a *vario-coupler* you will require a cardboard tube or a fibre tube about four and one-half inches in diameter and five inches long. A rotor form that may be purchased for a dollar or so. A quantity of No. 26 and No. 28 B. & S. gauge, double cotton covered magnet wire and some No. 20 of the same type wire. You will also require a brass shaft or rod 1/4 inch diameter, a dial, knobs, switch, a panel of fibre or bakelite about 3-16 inch thick and 6 inches square and a wooden or fibre base 1/2 inch thick and 6 inches square, besides screws, odds and ends, etc. The cardboard tube and rotor form are prefer-

HOW TO MAKE AND USE IT

ably soaked in paraffine but this is not essential. Begin winding the cardboard tube, starting ½ inch from one end through a small hole, as shown, and wind on 38 turns of the No. 26 wire, keeping the turns close together but not touching. Then, bring the wire across the tube as shown in *Fig. 37,* leaving a space of one inch bare and continue to wind on another 30 turns of wire. In winding, take off three taps from each section, one at every twelve turns, beginning at second turn from top and leaving two turns at bottom of winding as shown in *Fig. 37, T, T, T,* finally passing the end through a hole as shown. The whole should then be covered with paraffine, or it may be shellacked, although shellac will decrease its efficiency and paraffine will serve every purpose. In the centre, at the bare space which has been left, a ¼ inch hole should be bored as shown in the cut. The next step is to wind the rotor form, which is done by winding on twenty-five turns of No. 20 wire (starting through a fine hole as shown) which forms the *"tickler" coil Fig. 38, A,* and leaving the two ends of the wire about 6 inches long and running the last end through a hole to hold it. Then, at the other end of the rotor, wind on forty-two turns of No. 28 wire to form the secondary *Fig. 38, B,* and leaving free ends of wire about 6 inches long. Through the centre of the rotor-form, a hole should be bored ¼ inch in diameter as shown. Like the other coil, the rotor may be coated with paraffine or shellac, the former being preferable, and the ends of the coils may be first fastened by glue or sealing wax, to prevent loosening or unwinding.

The next step is to mount the coupler, which is done as follows: In the fibre panel, bore a ¼ inch hole two inches from the top and 3¼ inches from one side, *Fig. 39.* In the lower corner, place a switch with six contact points *A,* and on the right hand side drill six 3-16 inch holes for binding posts *B.* Then make and place the shaft in the rotor, securing it by glue, sealing wax or by means of nuts, according to your mechanical ability, and attach the two parts to the panel, fitting a dial and knob to shaft and mounting the panel on the base. In connecting up, the primary, secondary and tickler wires are connected to the six binding posts and the six tap-offs on the primary coil are connected to the switch contacts as shown in the cut, with one of the primary wires in the switch post. Then, when the coupler is to be set up, the aerial is connected to one of the primary posts, the ground to the other, the secondary posts are connected with the grid circuit and the tickler posts to the plate and receiver circuits, all of which is shown in the diagram *Fig. 40*.

A *variometer* may be made in a very similar manner, using two cardboard tubes, one about four inches in diameter and three inches long; the other three inches long and about three and three-quarters inches in diameter. The dimensions should be such that the small tube can turn freely, without touching, within the larger tube and the smaller the space between the two the better; but you must remember to allow for the thickness of the wire to be wound upon the inner tube. First, measure carefully the exact centres, so that when a shaft fastened to the inner tube or rotor is passed through the larger tube, the inner one will rotate freely and evenly without touching or increasing the space. The entire efficiency of the variometer depends very largely upon the accuracy with which this is done. Starting with a small hole about ¼ inch from the outer edge of the smaller tube, wind on about twenty turns of No. 24 double-coated, cotton-insulated copper wire, being careful to keep the turns separated. Then skip a space of about an inch, as shown in *Fig. 41*, and wind on another twenty turns, finally running the wire through a hole, fastening both ends with a drop of glue or sealing wax and leaving five or six inches of free wire at each end. Starting the same way, wind the larger tube in exactly the same manner and being sure to wind in the same direction. When all are wound, mount the smaller tube on a shaft inside of the larger tube, fastening shaft by glue or sealing wax dropped on from inside, and mount as shown in *Fig. 42*. Finally, connect one end of the stator wire to one end of the rotor wire, leaving plenty of free wire to allow rotor to revolve, and lead the other two ends to binding posts, as shown, being sure to keep that to the

HOW TO MAKE AND USE IT

rotor loose to allow free movement. The shaft to rotor should be equipped with knob and dial as shown in the cut and the whole mounted on a fibre or bakelite panel on a proper base. Do *not* shellac coils after winding as this will impair efficiency. If the winding has been properly done and the ends of wires fastened at holes by glue or sealing wax, there will be little danger of their becoming loose, but to further safeguard it, the coils may be covered with paraffine wax if desired. You must not expect this variometer to work as well as those purchased ready-made, for the efficiency of an instrument of this type depends upon the distance between rotor and stator, the least distance the better, but it will serve and will do very well for experimental use.

A SIMPLE CRYSTAL DETECTOR RECEIVING SET

THE accompanying diagram, *Fig. 43,* shows a very simple and effective little set which, under favorable conditions and with a good aerial, will pick up the broadcasted music, signals and other sounds from stations at considerable distances. No specific range for this or any other set can be given, for the efficiency of any set depends upon a great many conditions and influences. The length and height of aerials, the proximity of high buildings, electrically charged cables or wires, the perfection of insulation of aerial, adjustment of instruments; all affect the range of a receiving set and, in addition, there are climatic and other conditions to be taken into consideration. Moreover, similar instruments under similar conditions vary immensely and for no known reason according to locality, and an instrument which picks up sounds

HOW TO MAKE AND USE IT

from a certain station in one section of a city may fail utterly if installed a few blocks distant. As a rule, however, instruments in outlying country districts have a greater range and receive messages more clearly than those in cities, even though the latter may be much nearer the sending station.

In the accompanying diagram A represents the aerial, which should be a single wire as nearly 150 feet in length as possible and as high as it can be placed. B is the ground which should be made by scraping a spot on a gas, water or radiator pipe and soldering the wire in place. C is a variable condenser in the ground lead, and for this set should be about .0005 microfarads. D is the crystal detector, E the telephone head set and F a variometer.

The variable condenser and variometer may be purchased ready-made from any dealer in radio supplies and while they may be made at home yet it is far more satisfactory and just about as cheap to purchase the stock instruments. The same is true of the crystal detector and head set. You should, however, be careful in selecting the galena crystal to be used with such an outfit as this mineral varies greatly in its sensitiveness. The best plan is to purchase a pound or two of the crystals and test a number of pieces by means of a buzzer. You will probably find that while some crystals are absolutely useless others are fair and a few are very sensitive. To make up such a set is very simple, as the various parts are merely connected with insulated copper wire, as shown in the figure, using binding posts which may be purchased for a few cents. The whole may then be mounted on a piece of fibre-board or bakelite or it may be set up on a neat board or block and enclosed in a case with a hinged cover. Such a set, including all connections, wires, insulators for aerial, etc., should not cost over $20, and will be found far superior to many ready-made sets costing much more.

CRYSTAL DETECTORS AND HOW TO MAKE ONE

THE crystal detectors used in sets such as described are of various forms, *Fig. 44*, but in all, the principle is the same and they all consist of a crystal cup or holder, binding-screws and an adjustable contact of fine wire. They are not expensive instruments and it is usually easier and cheaper to purchase them ready made than to make them, but they are very easy to construct and any boy can make a practical detector in a few hours. One of the simplest is shown in *Fig. 45*, and consists merely of a fibre base; a strip of brass about 1-16 or ⅛ inch thick bent in the form shown; a brass plate which can be moved from side to side on a pivot, to hold the crystal; a fine coiled wire and binding-posts. The plate holding the crystal is connected with one post and the brass strip holding the wire to the other post. If possible, use platinum wire for the contact, but this is not essential.

LOOSE-COUPLED COILS

COILS, or as they are more often called, *tuning coils,* are very essential parts of radio telegraphy and telephony. The old style tuning coil, as used in wireless telegraphy, has been largely superseded by the type known as *loose-coupled coils* or *adjustable-coils* which may be altered or adjusted to tune much finer or more closely than by the old type coil. Although it is not difficult to make a loose-coupled coil yet, as is the case with many of the parts of radio sets, it is as cheap and far more satisfactory to purchase them ready made. The conventional type of loose-coupled coil consists of two distinct coils, one within the other, as shown in *Fig. 46.* One of these is the primary coil, the other the secondary or induction coil. The two are so arranged that the inner or secondary coil slips back and forth within the larger or primary coil, thus varying the coupling or induction, for the electricity—or oscillations—in the secondary coil is merely induced by the primary circuit in the outer coil, so that if a portion of the secondary coil is withdrawn from the primary coil, as shown in the figure, there will be less induced current and in this way tuning is accomplished. To allow of still finer adjustment, the primary coil is provided with an adjustable slider *A*, and the secondary coil has a multi-pointed switch *B*.

Another type of loose-coupled coil is arranged so that one coil revolves within the other; while another type, which is the simplest of all for the amateur to construct and gives the best results, is composed of three discs or coils *"stagger-wound"* which may be adjusted back and forth. To make one of these inductors you will require some stiff, smooth cardboard, heavy Bristol board, thin fibreboard or similar composition and about half a pound of No. 24 D. C. C. wire. Also, in setting up and arranging the coils, you will require binding posts, knobs, a little sheet brass and a few other odds and ends. With a pair of dividers or compasses draw three circles on the cardboard or fibre, each about four to five inches in diameter, having all exactly the same size. Then, using the dividers, scribe off an unequal number, five, seven, or nine marks around the circumference of each circle. *Fig. 47, A.* Next, still using the dividers, draw a smaller circle, say one and one-half inches to two and one-half inches in diameter within each circle (*B*). If the circles are four inches in diameter use the smaller circle inside, if five inches the larger one, and with a rule, draw radiating lines one-fourth of an inch apart from each of the marks on the outer circumference to the centre of the circle (*C*). With a pair of scissors, or a sharp knife (if cardboard is used) or a fine saw (if fibre), cut out the discs and cut slots in each disc according to the marks, as shown at (*D*).

Next, if you have used cardboard, give each slotted disc a thorough covering with shellac, using at least three coats, and when thoroughly dry proceed to wind the discs or coils. In doing this, start the wire—being sure to leave enough for connections—at a point at the inner end of one slot and wind over one segment and under the next, and as the number is uneven you will find that the wires will thus cross, as shown at (*E*). The number of times the wire should be passed can only be decided upon by experimenting after the coil is in use, but, as a starter, about twenty-five or thirty turns on one, about one and one-half times as many, or say thirty-eight to forty-five on the second, and twice as many on the third as on the first, or from fifty to sixty, will be somewhere near right. Then, by removing or adding a few turns, as you adjust your receivers you can finally secure the very best results. To mount this coil so it may be used, the coil or disc with the least turns, or in other words the *primary coil,* should be mounted rigidly and immovably and should be connected by means of binding posts to the aerial and ground wires. The secondary coil and the *tickler coil* should then be fastened to brass or metal strips about two to two and one-half inches long, one-sixteenth inch thick and half an inch wide. One end of each strip should be attached by small bolts or screws to the coils and the other end attached to a movable peg or bolt with a fibre or bakelite knob at the opposite end. *Fig. 48* shows clearly how this is done. In this way, the *secondary* and *tickler* may be swung back and forth to cover more or less of the primary; but great care should be used that the *tickler does not touch the primary coil.* When the coil is thus mounted on a proper panel or stand, it should be wired as shown in *Fig. 49,* and when the whole set is in good working order and final adjustments made, it should all be enclosed in a neat wooden case with a hinged top or cover, *Fig. 50,* although, of course, this is merely a protective measure and does not affect the working efficiency of the set. It must be clearly understood, however, that this type of coil can *only be used in connection with a vacuum-tube outfit as shown.* When all wiring is complete and adjustments are ready to be made, connect the storage battery, as shown; place the lamp or tube in its socket and gradually turn on the rheostat to see if the tube glows properly. *NEVER turn on the current to the bulb quickly or to full power, or the filament will be needlessly burnt out and wasted long before its time.* Next, connect the *B.* or dry battery, the ground and aerial and finally, the phones, which should be 2000 ohm receivers.

HOW TO MAKE AND USE IT

In using this outfit, turn on the bulb *slowly,* adjust the knobs carrying the coils so that all three are in line and then adjust or tune the variable condenser until the signals you wish to hear are clear. Then, by gradually adjusting the movable coils, you can cut out interference and also make the signals, music or other sounds louder. When the two movable coils are entirely away from the *primary-coil* you will have your shortest wave length while, when all three are together, you will have the longest wave length. So, by marking the knobs, you can always adjust these coils to receive any desired station's messages. If the coils cannot be adjusted to long enough or short enough wave lengths you can take off or put on a few turns of the wire on the discs. Also, *REMEMBER TO USE FLEXIBLE WIRE WHEN CONNECTING COILS TO TERMINALS IN ORDER TO ALLOW FOR PLAY AND MOVEMENT.* Such a set will give excellent results and by the use of amplifiers it may be used to reproduce sounds from stations loudly enough to be heard throughout a room.

AMPLIFIERS

AMPLIFIERS, which are devices designed to increase the volume of sound, are very simple affairs and while they may be bought ready-made any person can construct one in a few hours at much less cost. The most expensive part of an amplifier is the bulb or *vacuum-tube* and as such tubes as Myers are much cheaper than others and are exactly as good for amplification, though less efficient in the sending set—money can be saved by using them. A one step amplifier, on such a set as has been described above, will greatly increase the clearness and loudness of messages and by adding two or more, and using a loud tone horn or phone, music, singing, etc., can be produced to fill a large room or hall. In using two or more steps of amplifiers it is merely necessary to duplicate the first, which is shown in *Fig. 52*. The only thing to be borne in mind is *ALWAYS TO TURN THE TRANSFORMERS*

HOW TO MAKE AND USE IT

AT RIGHT ANGLES TO EACH OTHER FOR EACH STEP. This is essential in order to break up the magnetic fields and prevent the machine from squealing or roaring. Another point to bear in mind is that the *shorter the grid leads the more efficient the whole.*

In the figure, *A* is a *jack* or *plug-socket* which is connected as shown, the outer arms *a a* being connected back to the phone connections on the detector or original receiver set. Then, by inserting the plug of the phone between these (*a a*) the contact with *B B* is broken and you can listen-in as usual, while, by removing the plug and placing it at *C*, the message is amplified before you receive it. *Fig. 53.* Then, if a second step is used connect *c c* to the jack corresponding to *A* on the second amplifier, as at *b b*, and so on. The transformer *D* should be purchased ready-made and connected as shown and according to the directions accompanying it. In the figure, *D P* represents the primary terminals and *D S* the secondary terminals of the transformer and which will be found marked *P* and *S* on the transformer binding-posts.

E, in the illustration, is the grid. *F* the amplifying tube plate. *G,* the *B* batteries of 45 volts. *H* the storage battery, and *I* the rheostat. *Fig. 51.* The same six volt storage battery used on the detector set supplies the current for the amplifier and the connections to which are shown at *H.* Although a *grid-leak* is not essential to this amplifier one may be used if desired and should be shunted in as shown at *K.* One rule which you should always remember in setting up an amplifier or other radio apparatus is to *KEEP ALL WIRES AS SHORT AND DIRECT AS POSSIBLE for otherwise the set will howl and buzz.* Another thing is to *KEEP ALL WIRES FREE AND CLEAR AND WHERE THEY CROSS KEEP THEM WELL APART.* Also, *NEVER RUN TWO WIRES PARALLEL IF IT CAN BE AVOIDED AND IF IT CANNOT BE AVOIDED RUN THEM PARALLEL FOR AS SHORT A DISTANCE AS POSSIBLE.*

A SIMPLE VACUUM TUBE RECEIVING SET

THIS set is one of the simplest that can be devised to use a vacuum tube detector and it may be set up by anyone at a lower cost than the ordinary crystal sets may be purchased ready-made. Moreover, tuning in this set is simplified to the minimum, there being but two adjustments to be made, the variable condenser and the rheostat. The whole set is plainly shown in *Fig. 54*. *A* being the aerial, *B* the ground, *C* the variable condenser, *D* the inductance, *E* the grid-leak, *F* the fixed condenser, *G* the grid, *H* the plate of tube, *I* the rheostat, *J* the six volt battery, *K* the phone receivers and *L* the 22 volt B-battery.

In setting up this outfit, it is not necessary to purchase anything in the way of instruments with the exception of a variable condenser, the vacuum tube and socket, the rheostat, the receivers and the batteries and the entire set should be well within $30 in cost when completed.

The single inductance used is easily made by winding a pasteboard tube about three inches in diameter and two inches long with about forty-six turns of No. 26 double-covered, cotton-insulated copper wire. There should be a tap-off taken at the twenty-third turn (which is done by making a loop in the wire) and then the next twenty-three turns wound on. The fixed condenser and grid leak can be made as described in the chapter on *"condensers"* and the grid-leak, for best results, should be variable, which is easily arranged by using a paper and pencil leak and by adding lines or erasing them as described under "condensers." The fixed condenser should be about .001 mfd. and the variable condenser of from .0003 to .0007 mfds. Be sure to place the phone receivers and B-battery in series, the positive pole of the battery being connected to the tap-off on the inductance and the negative pole being led to the receivers as shown. In setting up and tuning you will very likely find it necessary to take one or two turns of wire from the upper end of the coil, which is easily done. In operating this set first bring the tube fila-

HOW TO MAKE AND USE IT

ment to a point where the oscillations produce a squeal or howl with the variable condenser set at zero. Then, by moving the condenser pointer over the scale slowly and carefully, you can determine the best point to receive signals. When this is determined, adjust the rheostat until the filament oscillates and then decrease the brightness a little. With a little practice and by marking the knobs you will readily be able to adjust the set very accurately and quickly.

REGENERATIVE RECEIVING SET

A REGENERATIVE set is one in which the wireless waves or oscillations entering from the aerial are *regenerated* and increased by means of the vacuum bulb and for this reason, as well as on account of others, it is far superior to any style of crystal set. Roughly, a vacuum bulb or tube detector is about thirty times as efficient as the best crystal detector and moreover, with such sets, one or more steps of amplification may be used, thus increasing the sounds to almost any extent. One of the great advantages of regenerative sets is that they will permit of tuning out interferences and the better the set and the more efficient the various units, the clearer and better the results obtained. I have already described how to set up a small regenerative set under "coils," but there are various other methods of arranging or assembling regenerative sets. One of

HOW TO MAKE AND USE IT

the best is the set used with a *vario-coupler* and *variometers* which is illustrated in *Fig. 55*. By using two variometers and a vario-coupler as shown, great selectivity of the circuit is obtained and by hooking on amplifiers and using a loud speaker phone, music, songs, speeches, etc., may be reproduced as loudly as a good phonograph. The only objection to such a set is that there are several adjustments to be made in tuning as there are the three knobs to look after. In making this set it is not advisable to attempt to make either the vario-coupler or the variometers, for while these instruments look simple, special tools and long practice and skill are required to construct them properly. Also, when purchasing these, select those in which the coils are not varnished or shellacked as these coverings invariably decrease the efficiency of a coil. The other parts required, aside from ordinary insulated electric bell wire, binding posts and a panel or base, are the *vacuum bulb* or *tube detector;* the *socket for the tube,* a *grid-leak and condenser* (which you can readily make yourself if desired), the *rheostat, a variable condenser of .0005 mfd.;* the *six volt storage battery;* *B-battery* and a pair of *2000 ohm receivers.*

By studying the diagram, you will easily see how the set should be arranged and hooked up, but if you desire, the wires may be altered to run at other angles or directions, the main thing being to keep them as short and direct as possible and to avoid running them parallel or crossing them. The *aerial A* is led in and connected to the primary of the *vario-coupler B,* the other terminal of the *vario-coupler* being connected with the *variable-condenser C,* and the latter being connected with the *ground-wire D.* From the *secondary coil* of the *vario-coupler E,* one wire is led to the *phone* or *receivers, F,* a connection being made at *G,* with the *storage battery H* and the *rheostat I.* The other wire from *E,* is led to the *grid-variometer J,* thence to the *grid-leak and condenser K,* then to the *vacuum-tube grid L.* The *plate-connection* of the *tube M,* is wired to the *plate-variometer N,* and from here the wire runs to the *B-Battery O,* the other terminal of which leads to the *phone-receivers, F.*

The whole affair, when finally con-

HOW TO MAKE AND USE IT

nected and mounted on a neat bakelite or fibre base, should be enclosed in a neat case with a hinged cover, in order to protect the instruments from dust and disturbance and yet allow inspection or adjustment. Of course the adjusting or tuning knobs and the outside connections for aerial, ground and batteries should be led through the case or box to binding posts on the exterior. If the whole set is arranged compactly and neatly and you make an attractive, well finished case, the set will be the equal in appearance, as well as in efficiency, of any ready-made set costing several times as much.

TRANSMISSION OR SENDING

I HAVE already described the difference between interrupted waves and continuous waves and how the human voice, or other sounds, if transmitted by the ordinary interrupted waves of a wireless telegraph sender, would be broken up and unintelligible. I have also explained how continuous waves are capable of carrying such sounds without breaking them up; but while sounds of voices, music, etc., cannot be properly transmitted over an interrupted wave yet code signals or alternate dots and dashes *may* be transmitted over continuous waves by means of instruments which break or interrupt such waves, so that a radiophone may be used to send both telegraphic and telephonic messages and a radio telephony receiver will receive both.

The principal item and most essential part of the sending or transmission apparatus, known as the radiophone, is the device by which the continuous waves are produced. The instrument which was first devised for this purpose was an ordinary *arc-light,* but this has now given place to a device known as a *vacuum-tube oscillator* which is similar to an ordinary incandescent light with specially prepared filament and other devices. The operation of the arc-light was as follows: If a condenser and inductance coil are shunted across the current to such a light, the current is lessened as the condenser is charged and the potential difference across the arc increased. This still further charges the condenser which discharges through the coil and again becomes charged in the reverse direction, the whole operation being repeated over and over again with extreme rapidity—as many as one million discharges a second being usual. Diagrammatically this arrangement is shown in *Fig. 56,* in which A represents the aerial, B the ground, C the coil, D the telephone receiver, E the condenser, F the light, G the dynamo or other source of electricity. The *vacuum-tube* on the other hand, acts in a very different manner. When this is charged with electricity the filament discharges an immense number of electrons upon a plate with incalculable rapidity *Fig. 57,* and this, by means of various appliances, produces the oscillating currents of extremely high frequency, or continuous waves. Unlike the arc light, moreover, the vacuum tube is employed in receiving, the incoming waves varying the current through the filament and so altering the flow of electrons, thus transforming the vacuum-tube from an *oscillator* to a *detector.*

It must not be supposed, however, that an oscillator, a condenser, a coil and an aerial and ground are all that are necessary in order to send wireless telephone messages. In addition, there are numerous appliances for tuning or adjusting wave lengths, modulating, amplifying and other purposes and in a large station these are very complicated and powerful. For example, the Westinghouse Station at Newark uses five 250 watt tubes—two being used as oscillators and three as modulators—and which work in conjunction with two fifty watt speech amplifiers. The current used in lighting is ten volt A. C. with five ampères of radiation. For sending the sounds, microphones are used and to maintain an even cool temperature on the tubes a rotary fan is employed. In this station the coil or inductance is a flat or pancake helix wound with half-inch copper ribbon while a special motor generator furnishes a 2000 volt current for the plate. The aerial is of the multiple-tuned type of six 150-foot wires at a height of 210 feet above the ground with a counterpoise of twelve wires on thirty foot spreaders twenty feet above the roof and below the aerial. As a lead-in, a ten wire cage on hoops two inches in diameter is used.

This, however, is one of the largest and most powerful stations in the world and for amateurs nothing so complete, so large or so powerful is ever required. For ordinary purposes, a very small sending set is all that is necessary and even for the smallest a license is required, for while there is no objection to any one operating a receiving set and listening-in to anything that is in the air, a multitude of signals and sounds being sent is a great nuisance and interferes with every legitimate sending station by confusing the sounds and creating interferences. In fact, the greatest trouble in receiving is interference and it is far more difficult to tune out the weak waves from nearby amateur sending instruments than to tune to the desired signals from the large stations.

Also, it is far more difficult to construct and set up a transmitting set than a receiving set and hence I advise all amateurs to leave the sending alone, or else make no attempt to install or use a transmitting set until thoroughly familiar with the subject and no longer an amateur. But as no book on wireless telephony would be considered complete without a description and instructions as to rigging up sending sets, a few simple directions and figures are given.

THE SIMPLEST SENDING SET

ALTHOUGH it may be possible to devise a transmission set or radiophone which is simpler than that shown in *Fig. 58*, it is questionable if anything simpler would really be efficient. At any rate, this is so extremely simple that the veriest amateur should have no difficulty in setting it up. It has only two adjustments, the rheostat and the variable condenser. Moreover, it is a very cheap set and, aside from the batteries, there is nothing which need be purchased ready-made (except wire) other than the following:

1 Variable condenser of .0005 mfds.
1 Microphone or telephone transmitter.
1 Rheostat.
1 Vacuum tube and socket.
1 Cardboard tube about two inches in diameter and about three inches long.

The diagram needs no explanation as to details; *A* being the aerial, *B* the ground, *C* the tap-off of inductance, *D* the variable condenser, *E* the B. battery of from 60 to 120 volts, *F* the six volt battery, *G* the rheostat, *H* the tube plate, *I* the grid, *J* the phone transmitter.

To make the inductance, wind the two inch cardboard tube with thirty-eight turns of No. 16 double cotton covered wire or B. & S. wire. When twenty turns have been taken, twist a loop in the wire as a tap-off and then continue winding the other eighteen turns. The tap-off should have the insulation scraped off in making the connection at *C,* after which the joint should be covered with adhesive tape. If, when using the set, any difficulty is experienced it may be tuned to a different wave length by taking off one or two turns of the wire at top or bottom, or both, of the inductance. For an aerial use No. 14 phosphor bronze 7 strand or No. 14 plain copper, using an aerial at least 150 feet long and of several wires and, if possible, use a counterpoise as described under *"Aerials."*

ANOTHER SIMPLE TRANSMISSION SET

ONE of the simplest transmission sets which can be devised for really practical work is that shown in *Fig. 59*. In this set, the only instruments which are required are as follows:

Vacuum tube with socket.

Variable condenser of .001 mfd.

Telephone transmitter or microphone.

60 volt B. battery

6 volt storage battery (ordinary dry batteries may be used).

2 Fixed condensers of .0005 mfd.

1 Modulation transformer or an old type Ford spark coil.

1 Rheostat.

In addition, you will need some No. 28 B. & S. double covered cotton insulated wire.

No. 26 B. & S. double covered cotton insulated wire.

2 pasteboard tubes ¾ inch in diameter and 2 inches long.

1 pasteboard tube 3 inches in diameter and 2 inches long.

A supply of ordinary cotton-covered bell wire.

Aerial wire (No. 14), seven-strand phosphor bronze or copper is best, but plain will do.

Also, to secure the best results, use a counterpoise as described under "*Aerials.*"

The diagram shows so plainly how this set is made that no detailed explanation is necessary. The inductance C is merely a single coil made by winding about 50 turns of the No. 26 wire on the pasteboard tube 3 inches in diameter. The radio choke-coil L is made by winding a few layers of the No. 28 wire on the pasteboard tubes ¾ inch in diameter and the modulation transformer N may be bought ready-made or an old style Ford spark plug with the buzzer or contact screwed down hard may be used. In setting up be sure the **PRIMARY COIL OF THE MODULATION TRANSFORMER** is connected with the *phone transmitter O*. One great advantage in this set is that there are only **TWO** adjustments to be made, the rheostat *H*, and the variable condenser *B*. In the diagram, *A* is the aerial, *B* the variable condenser, *C* the inductance or helix, *D* the ground, *E* fixed condenser, *F* tube plate, *G* grid, *H* rheostat, *I* 6 volt battery, *J* 60 volt battery, *K* fixed condenser, *L* radio choke, *M* grid leak, *N* modulation transformer, *O* phone transmitter.

AN EFFICIENT 5 WATT TRANSMITTER

FOR those who wish a more powerful and efficient set for sending this is to be highly recommended. Under favorable conditions it should have a range of from 25 to 40 miles. Moreover, it is not an expensive set to make and, exclusive of batteries, should not cost over $35.00 to $45.00. The accompanying diagram *Fig. 60,* makes the wiring very plain, *A* being the aerial, *B* the ground, *C* the ammeter, *D* the inductance, *E* the variable condenser, *F* the fixed condenser, *G* the grid-leak, *H* the modulation transformer, *I* the 6 volt battery, *J* the microphone, *K* the grid, *L* the plate, *M* the rheostat, *N* the choke-coil, *O* the rectifier, *P* the current transformer.

Although practically every part of this set can be made, with the exception of the ammeter, rheostat, microphone, and modulation transformer coil, still it is almost as cheap and much more satisfactory to purchase the current transformer, the variable condenser and the choke-coil. The inductance D is easily made, the plate coil being wound with No. 18 cotton insulated wire on a cardboard tube three and one-half inches in diameter, using twenty-six turns and is tapped at the thirteenth turn. The aerial coil is the same size of wire, but with only eight turns. The space between windings or turns should be about one-half inch. The choke-coil may be purchased as such, but an ordinary spark coil—using the secondary winding—may be used and an old Ford spark coil with contact-breaker screwed down may be used for the modulation transformer. The ammeter used should be one of low reading scale or, if desired, an ordinary 3 volt flashlight bulb may be used instead. The ammeter, however, will give far better results. The grid-leak is an ordinary lead pencil leak of medium soft lead. The most difficult part to make is the current rectifier, but even this is very simple and consists merely of eight pint fruit jars filled with a solution of ordinary borax in the proportion of half a pound of borax to ten pints of water. In filling the jars avoid having any undissolved borax or sediment in them, and fill only about three-fourths full. The plates consist of alternate lead and aluminum strips, eight of each alternating as shown in *Fig. 61*, and with each plate 5 x ¾ inches. With this rectifier and the current transformer *P*, an ordinary 110 volt, 60 cycle electric current may be used, or without these, two or three B-batteries may be used instead, but this is far more expensive and does not give as good results. Before using this outfit the rectifier must be treated to form the plates, which is accomplished by connecting an ordinary 50 watt incandescent bulb and letting the 110 volt current run through it for ten or twelve hours.

Another point to remember is *always* to disconnect the high voltage current from the plate when not in use and when using the set *always* light the filament in the tube with the low voltage battery before turning on the high voltage current. The switch on the microphone circuit also should *always* be thrown off, thus disconnecting the phone from the battery, when

HOW TO MAKE AND USE IT

not in use. To tune this set it is only necessary to adjust the variable condenser until the ammeter or flashlight shows the highest reading.

USEFUL THINGS TO REMEMBER

THAT a *crystal detector* set is *never* as efficient as a vacuum-tube set, no matter what sort of equipment you use.

That a *crystal detector* set *cannot* be satisfactorily amplified.

That most *small* or *cheap* sets *cannot* tune out local interferences.

That *grid-leaks* are *not* used on crystal detector sets, but only on vacuum-tube sets.

That a *large-tuning-coil* with wires spaced closely will give better results than a *short coil* or one with wires far apart.

That a *vario-condenser* and a *loose-coupler* gives finer tuning.

That a *vacuum-tube* is about *thirty times* as efficient as a crystal detector.

That a *wave length* has little to do with the distance you can receive.

That *3000 ohm receivers* will often raise a cheap set from inefficiency to excellency.

That a *loading-coil* is not needed with a *loose-coupler* and *variable-condenser*.

That if *two or more crystal sets* are used on one aerial only one can be used at one time and a switch must be provided to throw sets out and in.

That a *loose-coupler* is better than a *tuning-coil*.

That a *loose-coupler* should be placed between aerial and ground.

That *money saved* in buying cheap head phones or receivers is really thrown away and that a great deal depends upon the phones.

That the *fixed-condenser* should be shunted across head-set.

That in setting up *an aerial* one long wire is better than many shorter ones.

That the *lead-in* counts and a long lead-in is an advantage.

That *aerials* and *lead-ins* should be insulated from everything else.

That *aerials* should be placed as high as possible.

That when placing *aerial* near elevated structures, wires, bridges, or steel buildings it should be placed at right angles to them and as far away as possible.

THE HOME RADIO

That *7 strand, phosphor-bronze* wire is the best for aerials, but ordinary No. 14 copper wire will do.

That *continuous waves* penetrate everything.

That the *lead-in* from aerial should be at end of aerial which is *towards* the sending station you most often wish to hear.

That *aerial* does *not* have to be *horizontal*.

That for *sending*, a many-wire aerial is far better than a single wire.

That a *counterpoise* is better than ground, particularly in sending.

That an *indoor wire* will serve for an aerial, but is not so good.

That an *iron bedstead* or *spring-bed* will do for an aerial in case of necessity.

That the *simplest* and *cheapest vacuum-tube* receiving set is better than the best crystal set.

That the best form of receiving set is the *regenerative* set.

That a *vacuum-tube* or *regenerative* set may be amplified to almost any extent.

That an ordinary *phonograph horn* attached to a head telephone receiver will increase the sounds *somewhat* and *will* act as a loud speaker.

That a *variable-condenser* helps fine tuning.

That the *filament battery* of a vacuum-tube set may be a dry battery, but that it is more expensive in the end than a *storage battery*.

That the *Ultra Audion* circuit has the plate circuit led back to the honeycomb-coil and amounts to a regenerative set.

That the best type of vacuum-bulb receiving set is the *regenerative* with amplifiers.

That each step of *amplification* requires another tube.

That if there are too many turns on the *inductance* they may be taken off to secure tuning.

That the *distance* you can receive depends upon various climatic and other conditions.

That the *filament lighting* does not always mean the set is operating properly.

That if *filament rheostat* is turned on suddenly the *filament* may be paralyzed and must be left to recuperate before it will glow.

That *burning the filament* too brightly merely wastes the filament and shortens

THE HOME RADIO

the life of the tube without adding to the efficiency of set.

That a *variable grid-leak* can be made with pencil marks on paper and may be altered by erasing or adding lines.

That *some tubes* are best for *detectors,* others for *amplification* and others for *transmitting* or as *oscillators.*

That it is *often cheaper* to buy ready-made accessories than to make them.

That *all joints* in wires (except in binding-posts) should be soldered.

That the *ground connection* should be soldered to a water, gas or similar pipe or to a large copper plate buried in the ground.

That the *steel girder* or frame of a building makes a good ground.

That *sending or transmitting* sets must have a license to comply with the law.

That the *fire departments* have special regulations regarding the installation of aerials.

That an aerial cannot be placed *across a street* without permission.

That aerials do not *attract* lightning and if provided with a gap or *lightning switch* are perfectly safe.

That the *best* in the way of materials is always cheapest in the end.

That when using a *sending* set the *low voltage* should be turned on first or the tube may be ruined.

That in a *sending set the battery* should *always* be turned off from the phone circuit when not in use.

That a *sending set* is always better with a *counterpoise* than with a ground.

That it does *not* pay to try to make certain instruments.

That while *wireless telephones* are so easily adjusted and simple a child may use them, they are also very delicate affairs and are easily put out of adjustment or ruined by carelessness.

That you must *not* expect too much for your money in ready-made or home-made sets.

That a set may *act very differently* on different days or under different conditions.

That you should *not condemn* your instruments until you are sure the fault is not in yourself.

That *loose connections*, poor insulation, poor ground, poor joints in wires, worn insulation, wires crossing and many other small matters may put a set completely out of business.

That you can *seldom improve* upon a ready-made set by adding anything to it, but can do better by building a new set.

That every *accessory or piece* of apparatus is made for a specific purpose and that you should consult the manufacturers or dealers as to the best for your purpose before purchasing.

That the most *expensive sets* are not always the best, as oftentimes finish, cabinets and elaborate fittings add to cost without increasing efficiency.

That while a *receiving set* may be made to go inside a *safety match box* such things are merely toys and are not for household use.

That when a *dealer advertises* that a cheap set can receive signals from a certain distance, be sure to find out if he means *code signals* from *radio telegraphic* stations or sounds of voices, music, etc. *No one* can *guarantee* how far a set will receive as too many outside factors influence this.

HOW TO MAKE AND USE IT

That like *everything else* each and every maker claims his sets are the *best*. Investigate *several* before buying.

That *anyone* with the least mechanical ability can build wireless telephone sets if they purchase the parts which require special knowledge, skill or devices for making.

That the *prices* of most sets do *not include batteries, tubes* or *phones*.

That a *storage battery must be recharged* as soon as it becomes weak or your set will not work.

That the *vacuum-tube* is one of the most delicate devices ever invented and should be treated accordingly.

That *no license* is required for receiving sets and the air is free to all who want to listen-in.

That all *broadcasting* stations publish their daily programmes.

That the *worst interferences* are the nearby sending stations. So don't add to others' troubles by sending unless you have *good reasons* or are *sincere in your experiments*.

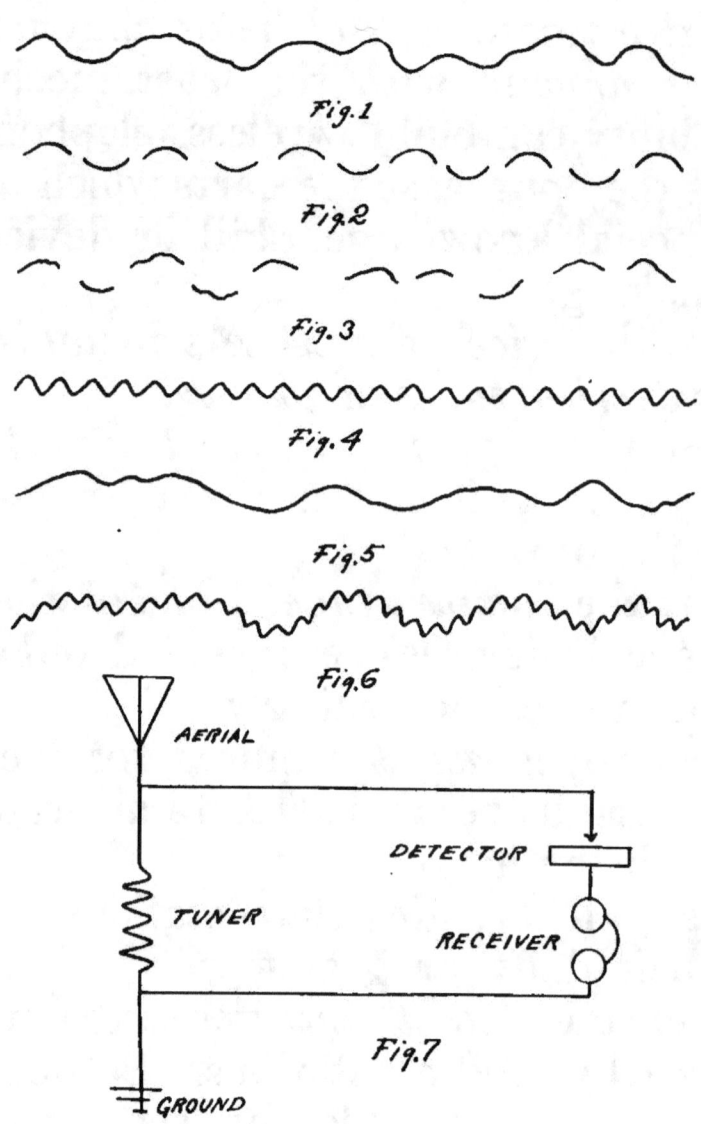

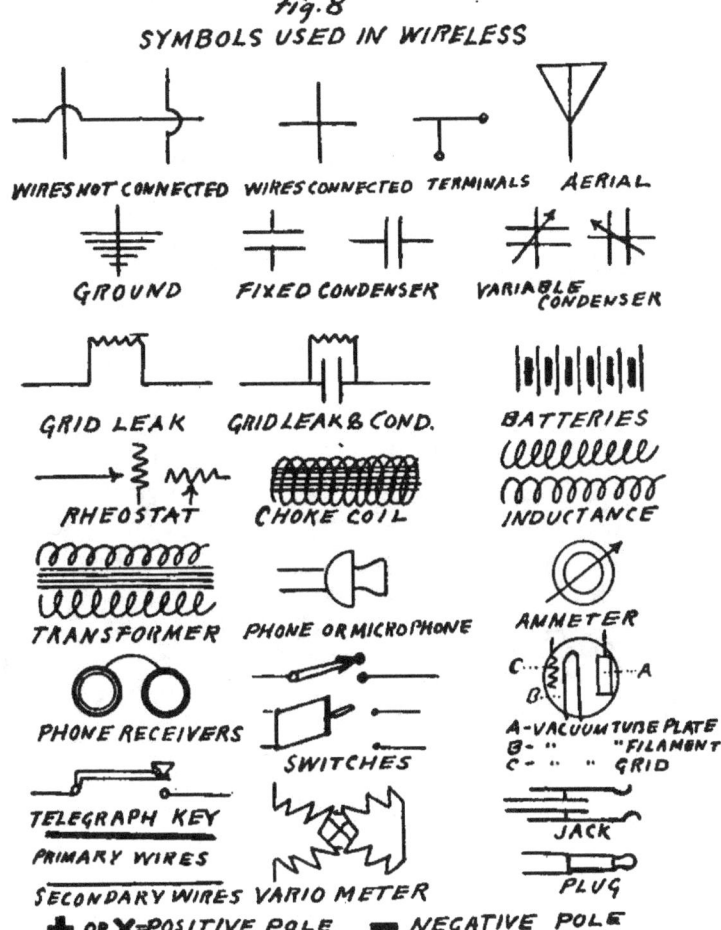

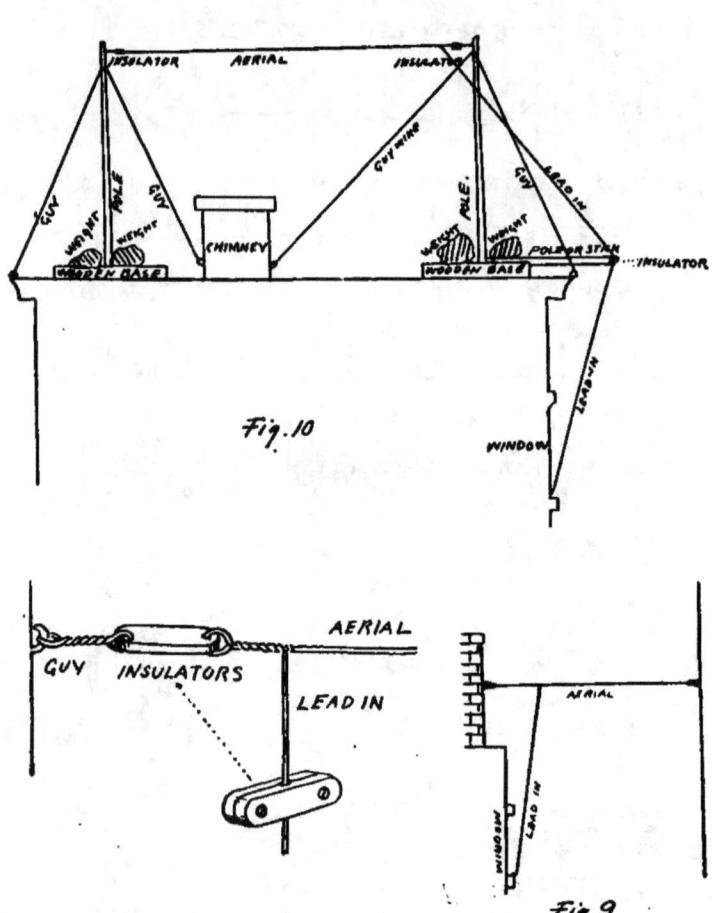

Fig. 10

Fig. 9

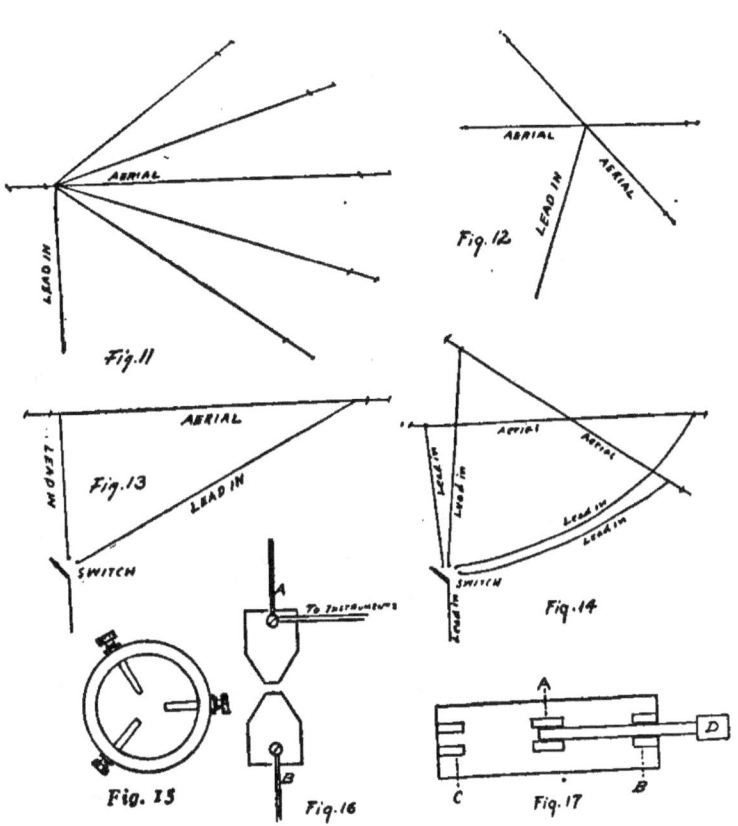

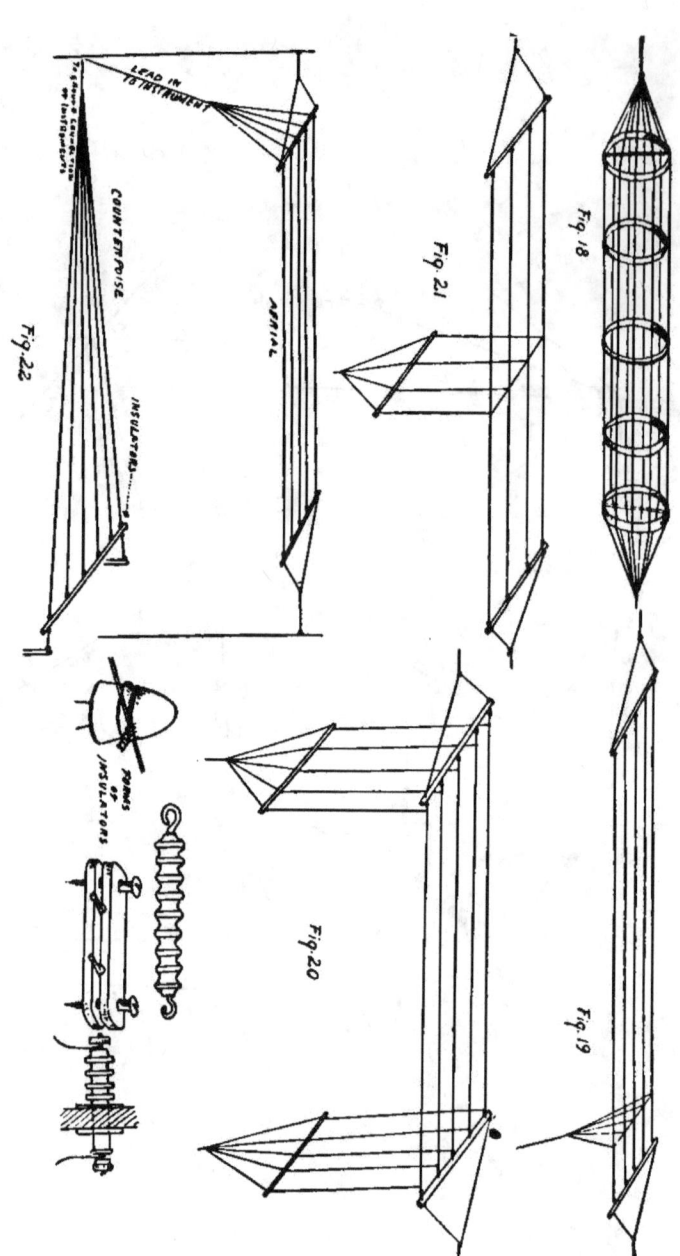

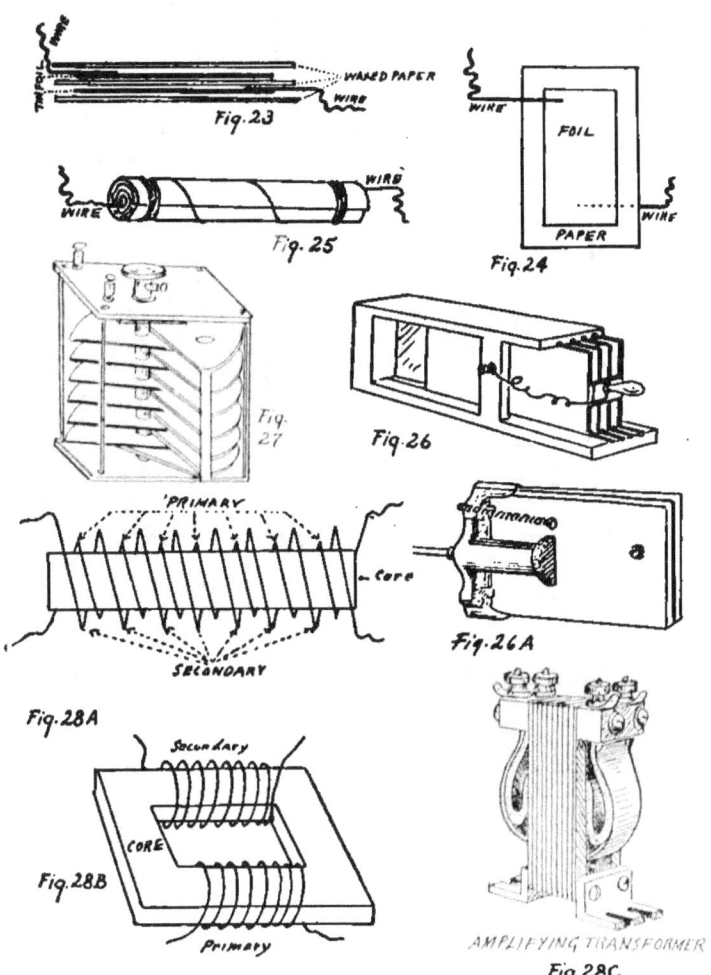

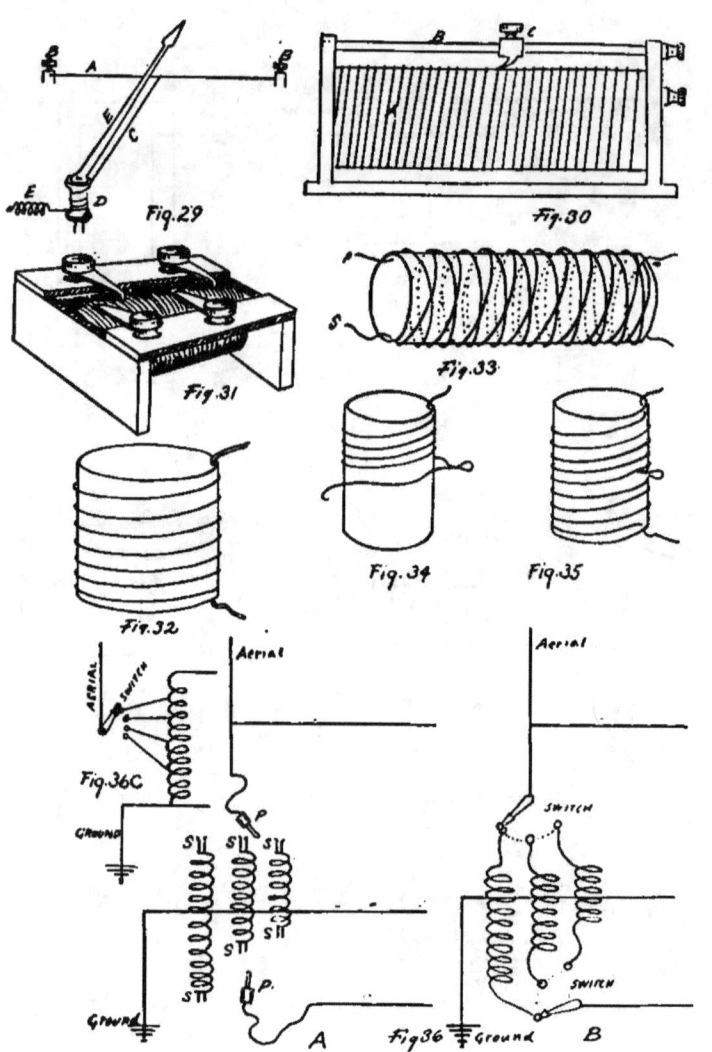

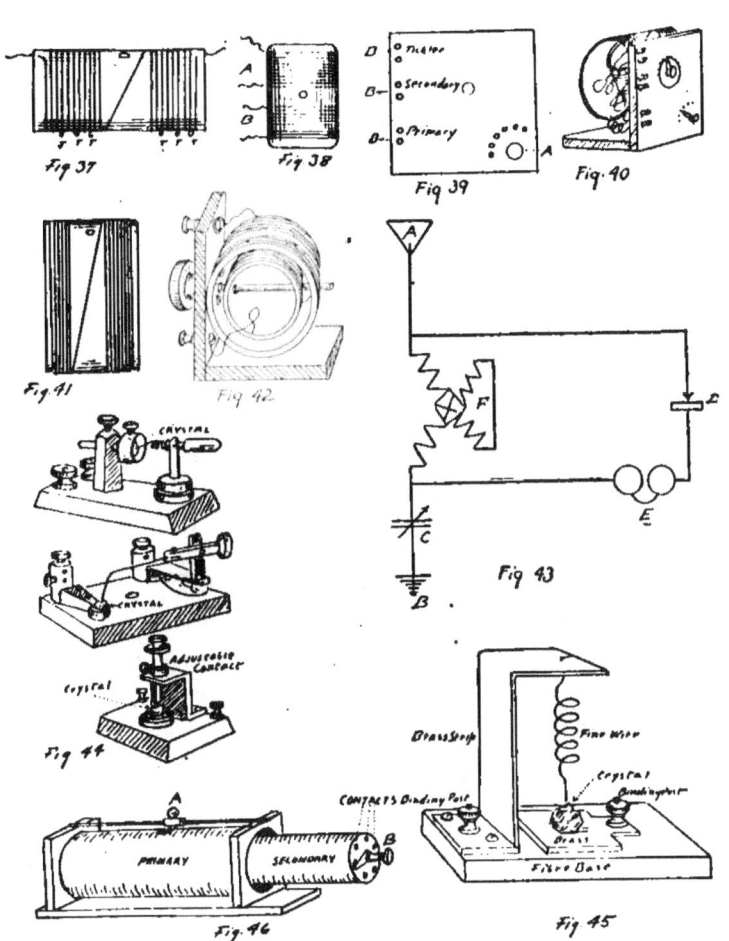

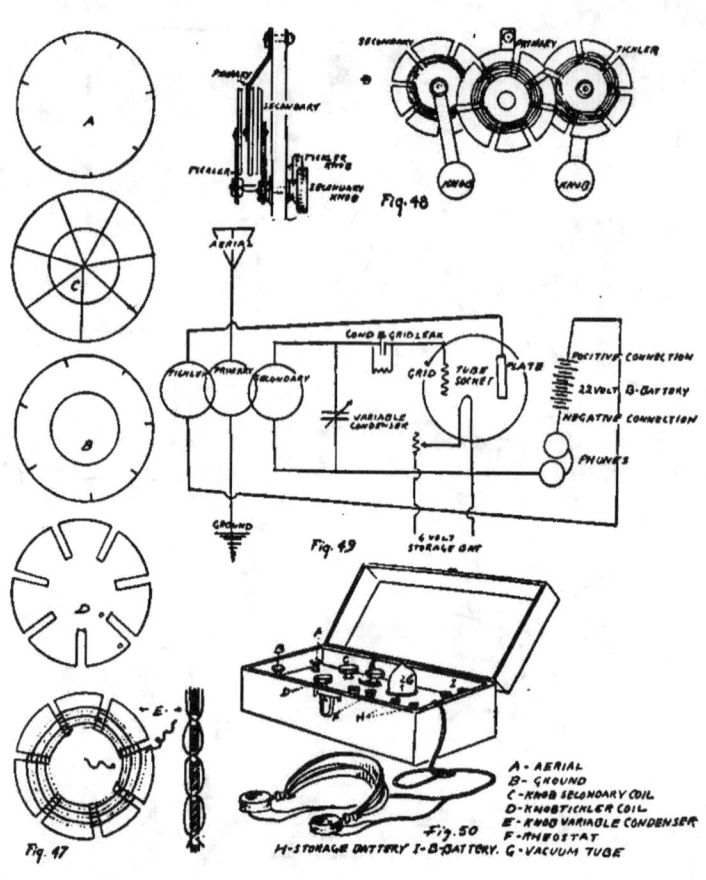

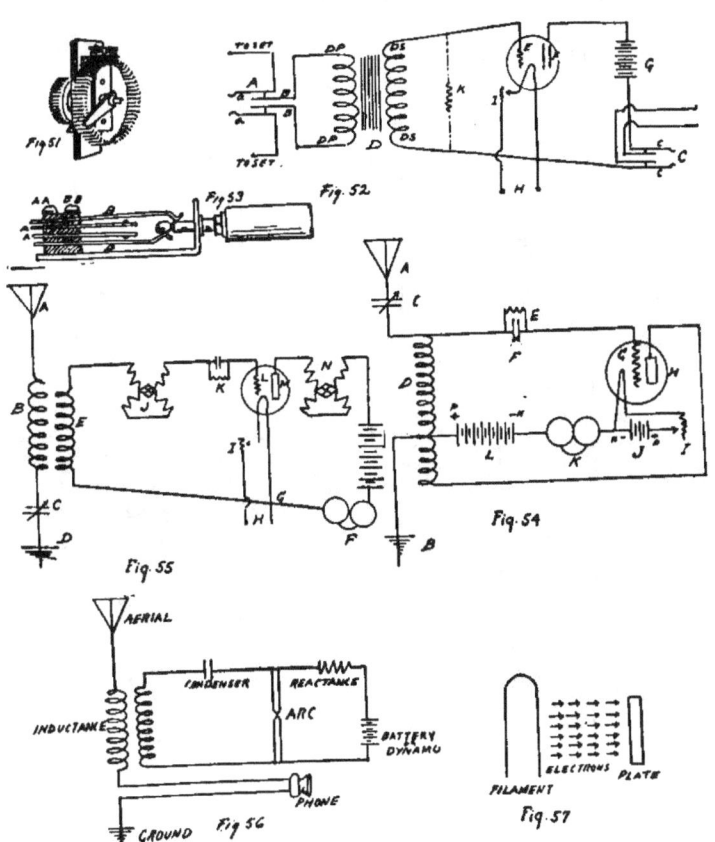

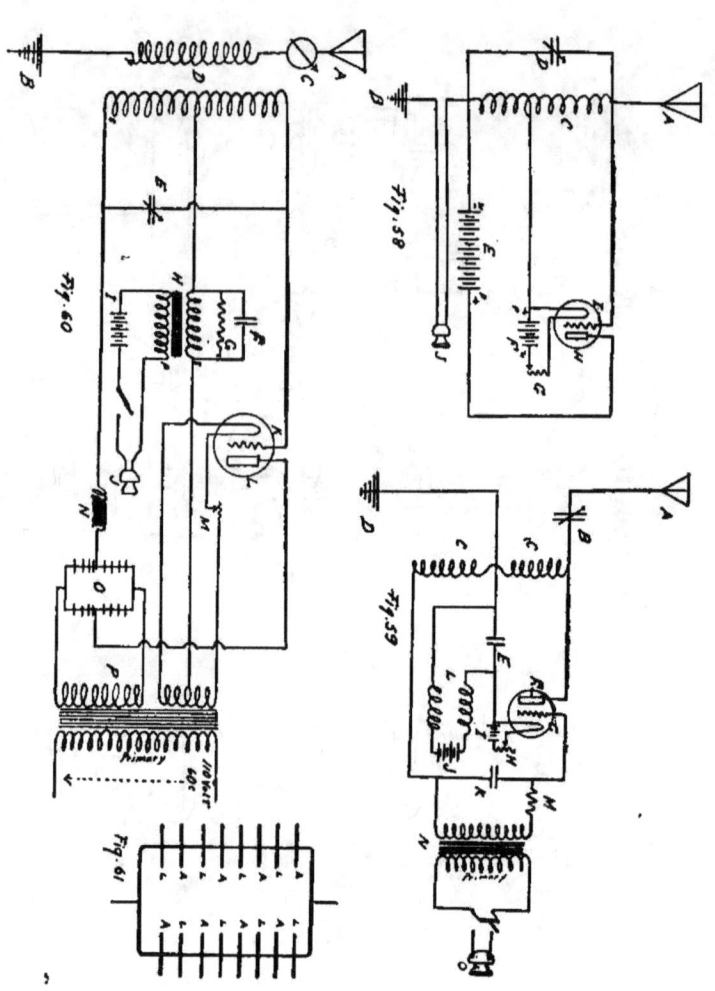

LIST OF DIAGRAMS

	Fig.
Variation in vibration of word	1
Interrupted waves	2
Word broken by interrupted waves	3
Continuous waves	4
Sound wave of word	5
Word carried on continuous wave	6
Units of receiving set	7
Symbols used in diagrams	8
Aerial on roof	9
Aerial on posts	10
Directive aerial	11
Directive aerial	12
Directive aerial	13
Directive aerial	14
Anchor-gap	15
Air-gap	16
Lightning switch	17
Cage aerial	18
Transmission aerial	19
Transmission aerial	20
Transmission aerial	21
Counterpoise	22
Tin-foil condenser	23
Making tin-foil condenser	24
Condenser completed	25
Plate variable-condensers	26
Rotary variable-condenser	27
Transformers: A—Open Circuit. B—Closed core. C—Amplifying	28
Hot-wire ammeter	29

LIST OF DIAGRAMS

	Fig.
Tuning-coil with slider	30
Tuning-coil with contacts	31
Simple coil or helix	32
Inductance coil	33
Coil with tap-off	34
Coil with tap-off complete	35
Coils plugged out and in	36
Winding primary for vario-coupler	37
Rotor primary for vario-coupler	38
Panel for vario-coupler	39
Vario-coupler mounted	40
Winding variometer	41
Variometer mounted	42
Simple crystal receiving set	43
Crystal detectors	44
Home-made crystal detector	45
Loose-coupled coil	46
How to make a staggered inductance	47
Inductances set up	48
Wiring simple regenerative set	49
Regenerative set in case	50
Rheostat	51
One-step amplifier	52
Jack and plug	53
Simple vacuum tube set	54
Vacuum tube set with vario-coupler	55
Arc light transmission system	56
Vacuum tube electrons	57
Simplest transmission set	58
Another simple transmission set	59
A 5-watt transmission set	60
Details of rectifier	61

BOOKS ON ELECTRICITY

A-B-C OF ELECTRICITY
By William H. Meadowcroft

Simple and clear explanations of the various ways by which electricity is obtained and applied.

HARPER'S BEGINNING ELECTRICITY
By Don Cameron Shafer

A secondary book containing general explanations and their applications, with instructions in making simple electrical appliances.

HARPER'S EVERY-DAY ELECTRICITY
By Don Cameron Shafer

How to use electricity with explanations of practical ways and means for electrical experiments and home construction.

HARPER'S ELECTRICITY BOOK
By Joseph H. Adams

A comprehensive, practical book showing how those who are interested in electricity can carry their experiments further and gain working results for themselves in a wide variety of fields.

HARPER'S HOW TO UNDERSTAND ELECTRICAL WORK
By William H. Onken, Jr. and Joseph B. Baker

A general guide to a knowledge of the great commercial applications of electricity with which we are surrounded—clean-cut explanations of the modern uses of electricity in transportation, manufacturing, farming, mining, the household, etc. A book necessary to a knowledge of the work of modern life.

HARPER & BROTHERS
NEW YORK   Established 1817   LONDON

HARPER'S PRACTICAL BOOKS

HARPER'S BOOK FOR YOUNG GARDENERS
HARPER'S INDOOR BOOKS FOR BOYS
HARPER'S OUTDOOR BOOK FOR BOYS
HARPER'S CAMPING AND SCOUTING
HARPER'S BOATING BOOK FOR BOYS
HARPER'S ELECTRICITY BOOK FOR BOYS
HARPER'S BOOK FOR YOUNG NATURALISTS
HARPER'S HOW TO UNDERSTAND ELECTRICAL WORK
HARPER'S MACHINERY BOOK FOR BOYS
HARPER'S HANDY-BOOK FOR GIRLS
THE STORY OF OUR GREAT INVENTIONS
MOTOR-BOATING FOR BOYS

Each Volume Fully Illustrated.   Crown 8vo

HARPER & BROTHERS
NEW YORK   Established 1817   LONDON

www.ingramcontent.com/pod-product-compliance
Lightning Source LLC
Chambersburg PA
CBHW062354220526
45472CB00008B/1799